ISO 14001
Environmental Systems
Handbook

ISO 14001 Environmental Systems Handbook

Ken Whitelaw

Butterworth-Heinemann Ltd
Linacre House, Jordan Hill, Oxford OX2 8DP
A division of Reed Educational and Professional Publishing Ltd

 A member of the Reed Elsevier plc group

OXFORD BOSTON JOHANNESBURG
MELBOURNE NEW DELHI SINGAPORE

First published 1997

British Library Cataloguing in Publication Data
A catalogue record for this book is
available from the British Library

Library of Congress Cataloguing in Publication Data
A catalogue record for this book is
available from the Library of Congress

ISBN 0 7506 3766 8

Typeset by Butford Technical Publishing
Bodenham, Hereford
Printed and bound in Great Britain by
Biddles Ltd, Guildford and King's Lynn

Contents

Preface

What is an environmental management system?

All organizations have some impact on the natural environment, particularly through the resources they use, the processes and activities they undertake, and the waste they create. However, many organizations do not actively seek ways of reducing these impacts.

Each business, of course, has its own unique set of impacts on the environment and a limit as to how much improvement it can make to its environmental performance. But the fact is that every business, no matter how small, can achieve results if improvements are planned and carried out in a structured manner.

Environmental management is the process whereby organizations assess, in a methodical way, the impacts of their activities on the natural environment, and take action to minimize these impacts. An Environmental Management System is a management system that allows an organization to control its environmental impacts and reduce such impacts continuously.

What is ISO 14001?

In essence, an environmental management system that is designed to achieve what has been outlined above, will be a system that will meet the requirements of ISO 14001:1996 – the International Standard for Environmental Management Systems.

That is the ISO 14001 Standard defined in its most simplistic of terms. Of course, there is considerably more to it than that. For example, the organization must have a management system that identifies its environmental impacts and ensures that it complies (as a minimum) with current environmental legislation. Commitment to continuous improvement must be demonstrated via the vehicle of objectives and targets being set, monitored and measured and the whole management process being subjected to internal scrutiny and review. A publicly available environmental policy must also be produced which reflects the above activities.

ISO 14001:1996 was first published in September 1996 and so is a very young Standard – although international agreement was reached to allow certification against the draft standard DIS/ISO 14001 from December 1995. The speed of development to this stage has been remarkable compared to quality assurance standards: for example, the ISO 9000 series of standards – ISO 9001, ISO 9002 and ISO 9003 – which had operated over a period of some 30 years in various guises before finally becoming full ISO standards in 1987.

ISO 14001:1996 was an environmental specification that was designed to be audited against by external, or third party, independent certification bodies. Therefore, the important process of such independent certification merits a separate chapter in this book.

Against expectations perhaps, ISO 14001:1996 does not set out absolute environmental performance guidelines. No maximum levels of volatile organic compounds (VOCs) emitted to atmosphere, no maximum volumes of effluent, no maximum tonnage of waste sent to landfill are quoted. The individuals and committees responsible for the drafting of the Standard, having had prior experience of writing BS 7750 (ISO 14001's precursor), recognized that every business is different; therefore to set or prescribe absolute levels would be too complex an undertaking, as well as being unworkable.

The intended readership

This book is intended to be read by those managers of an SME (small to medium enterprise) in either the manufacturing or service sector. Such a manager, owner or managing director may be under pressure from customers to generate an environmental policy, or even to implement a recognized environmental management system by a certain target date. Within such a small organization, it is unlikely that there will be the necessary expertise available. The smaller business is also unlikely to have the necessary resource of time to spend on the implementation phase, there being no 'slack' in the management structure inherent in smaller organizations. A learning process is required and attendance at seminars and training courses may be out of the question, due to the restrictions on time. The high costs of consultancy may not be a viable option and the support network available to a larger organization again may not be an option.

The reader may also be the quality manager of an SME who, due to customer pressure in the past, has implemented a quality assurance system and obtained certification to ISO 9000, and thus whose expertise has made him or her the natural choice by senior management to implement ISO 14001.

Whether this choice of individual is correct or not is irrelevant. The fact is that many such managers, given the task of implementation, are first and foremost from a quality assurance background. Thus, such a manager may be thrust into the environmental arena and may be required to operate in areas beyond their training, understanding or knowledge.

The managers of SMEs, as a target readership, have been chosen because SMEs have been neglected in some respects as potential readers in other environmental texts. The style and level of debate, discussion and information imparted in much of the environmental literature is aimed at the specialist manager from the larger organization. SMEs should not be forgotten as they are a powerful driving force in any economy: in the United Kingdom, even more so. They create jobs, are a source of innovation and competition, create a dynamic, healthy market economy and preserve a stable economic base. However, many SMEs are:

- Still unaware of relevant environmental legislation

- Unconvinced of the potential cost savings and market opportunities of environmental management systems

- Out of step with their customer requirements regarding environmental probity

- Dissociated from their stakeholders' concerns about the environment

They see no reason to address their environmental aspects and any action in this area is often a negative response to legislative and regulatory pressures, rather than an attempt to pro-actively seek new opportunities from managing potential environmental issues.

The intended purpose of the book

When attending seminars, or reading books and articles on environmental standards, it seems that they are very much aimed at the professional environmentalist and really not very accessible to the generic manager of a small organization. The needs of smaller organizations are not addressed and some pragmatic advice on the whole process of assessment and certification is missing. In particular, the following information is difficult to obtain from the available literature:

- The concepts of environmental management systems in easy-to-understand language

- The stages of the certification process

- An understanding of the individual external auditor's methodology

- Environmental auditing methodology

- Availability of guidance material

- Fusion of management systems (integration)

In fact, much of the literature available on the market focuses very much on the requirements of environmental legislation, a difficult and complex area, which possibly makes some SME managers feel that environmental management systems are too much of a specialist subject to contemplate. Even those SMEs who are not under any customer pressure to implement but wish to pro-actively address their environmental responsibilities, may have decided that it is beyond their capabilities and that they will wait until customers, for example, force the requirement of ISO 14001 upon them.

The purpose of the book, therefore, is to assist implementing organizations to gain ISO 14001 certification. Whilst the focus of the book is on the steps required for successful implementation, there is an additional emphasis on the assessment process as enacted and performed by an external body – the certification body – on the environmental management system of an organization.

For example, enough information is included in Chapter 3 to make sure that any organization implementing an environmental management system will build an appropriate and meaningful system suitable for their size, complexity and needs. However, the intention is not to replace either the Standard itself or the assistance that can be given by environmental consultants. A copy of ISO 14001:1996 should be purchased (the author saw little advantage to be gained by repeating the text of the Standard), and read in conjunction with the book. If an organization has any doubts as to the capability of its resources after such reading, then external environmental management systems consultants or environmental experts should be involved, to assist the implementation process.

The aim of the author is to facilitate the smooth progression of any organization towards successful certification. This is in everyone's interests and ultimately such sound environmental management practices contribute towards the wider environmental issues, such as global sustainability.

After reading the book, some of the misunderstandings that have developed (even in the short life of certified environmental management systems BS 7750 and ISO 14001) will hopefully have been corrected. The organization can then build a meaningful system, with or without consultants assisting in the process, depending on available resources. The emphasis on understanding the concepts is, of course, one way of correcting such misunderstandings.

International dimensions

The book is written for organizations operating within the UK – this being the author's native country and where most of the companies he has worked with are based. Appropriate legislative requirements, when referred to, and financial assistance schemes, are UK-based. Even so, the international dimensions of ISO 14001 are not ignored and, where possible and appropriate, the global issues are referred to. However, a book of this size cannot include all global environmental legislation and regulations, this being such a rapidly changing area, and even restricting such legislative controls to the UK is a difficult task. The reader is referred elsewhere in Appendix II for information on how to keep abreast in the UK. Certainly, reading the journals relevant to a particular business, as well as environmental journals or updating services – or even communicating with Government bodies – will assist in this process.

Legislation

National legislation and regulations are in a constant state of flux. Indeed, there are a number of agencies make their living by compiling such statutory information and supplying an updating service to subscribers, charging a fee for this service.

Compliance with legislation is a fundamental requirement of ISO 14001 and cannot be ignored in a book such as this, but there are better alternative mechanisms for organizations to expand their knowledge of appropriate legislation and keep that knowledge up to date: the agencies referred to above or the services of an environmental specialist or consultant.

Therefore, no attempt is made to reference the vast amount of legislative controls both in the UK and overseas, apart from suggesting some updating agencies in Appendix II. In the UK, process guidance notes and legislation documents can be bought from Government bookshops, although this can be a costly exercise if the organization is not sure which legislation is applicable or forthcoming. Again, if in doubt, use the services of an environmental specialist.

About the author

The author is an environmental management systems auditor, so it is inevitable that this viewpoint predominates some sections of the book. There are also many opinions expressed, as well as some ideas and examples, which are the author's alone and not necessarily the same as those of the Standard's authors.

Much of the inspiration for this book, some of the examples used, and some specific points discussed in depth, are based upon material presented by the author to a spectrum of interested parties, both in the UK and overseas, during workshops, seminars and lectures. These parties included:

- Potential clients

- Environmental managers

- Quality assurance consultants

- Quality assurance managers

- Business links organizations

- Local authorities

- Post-graduate MBA students

- Lending/insurance institutions

Much of the feedback from these presentations, and indeed many of the questions asked by the participants, have been used as a basis for the structure of the book. Some questions were of a recurring theme, perhaps reflecting that not enough publicity has been given by the certification 'industry' – the authors of the Standard, the accreditation bodies, the certification bodies and the environmental consultants – to give a clear and unambiguous message to business in general. This book is one way of addressing this shortfall in communication.

Author's assumptions

The author assumes that the reader is familiar with the concept of formalized, documented management systems – even if the exposure to such systems has been minimal, and no attempt is made to cover the basics of management control. A background knowledge of working with ISO 9000 quality assurance systems would be helpful but this is not essential. It is hoped, however, that the common sense viewpoint used throughout this book will ease any learning pains and encourage further reading on the subject.

Every undertaking has its own jargon – its own vocabulary – and where possible this is avoided; however, to assist understanding, key words and phrases are explained in Appendix I.

Order of reading

As with any book, it can be started at the beginning and followed through to the end!

However, the reader may be selective. For example, if only ISO 14001 certification is required and EMAS is of no interest, then Chapter 8 may be ignored. The following list contains a brief synopsis of the chapters, to allow the reader the choice of where to start.

Chapter 1

Start here to understand the background of environmental management in the wider sense, with environmental sustainability and environmental disasters being prime movers and triggers respectively. This chapter also shows how managing for the environment evolved, and the subsequent market demand for an environmental standard developed. A brief history of ISO 14001 is described – its birth as BS 7750 (the world's first environmental management system standard) – and where ISO 14001 is now: the international standard for environmental management systems.

Chapter 2

The reader is encouraged to read this chapter if an insight into the concepts and intended 'spirit' of environmental management systems is required. Without this fundamental understanding, an organization may well design a system that will meet the requirements of ISO 14001, but in doing so may create one which is an administrative burden rather than a tool used for improvement. The clauses of the Standard are described, but only in outline and purpose.

Certainly, the reader unfamiliar with formal documented management systems and one who finds the Standard itself a difficult document to read, particularly the detailed requirements contained in the sub-clauses, would do well to read this chapter before reading Chapter 3.

If the reader is a quality assurance professional, perhaps a manager or consultant, then reading this chapter is a good investment in time because there are fundamental differences in the concepts and philosophy between quality assurance standards and environmental management standards.

Chapter 3

If the reader is familiar with the concepts (and perhaps has had some awareness training), and wishes to prepare an organization for certification as soon as possible, then this chapter is the starting point.

This chapter discusses, clause by clause, sub-clause by sub-clause, what the Standard requires of the organization. Wherever possible, examples are given to illustrate approaches taken by other organizations, not particularly for the organization to copy, but to stimulate ideas and thus more easily meet the detailed requirements. Thorough reading and understanding of this chapter is essential prior to beginning the implementation process.

Chapter 4

There is a danger that because the organization is so busy implementing the Standard, it may neglect being prepared for the practicalities of the external assessment process. Clearly the organization needs to know what to expect from the certification body itself (who will be awarding the ISO 14001 certificate): its organizational structure, assessment methodology and the constraints and controls it has placed upon it.

Dealing with the certification body is part of the process of environmental improvement and should be seen as such. Fear and uncertainty do not make for good business, and so this chapter attempts to close any gaps in the understanding by the client of the certification body. It also seeks to encourage a co-operative approach – a business partnership with the client seeking certification – therefore allowing value to be added to the client's environmental management system.

Chapter 5

If the reader's organization has an existing quality assurance system and perhaps an occupational health and safety management system (or indeed other management systems), then this chapter will hopefully stimulate some thoughts and ideas on how best to combine or integrate these systems. Such integration should lead to improved efficiencies in managing the business generally. The fact is that all management systems have the same foundations; a common 'core' of manuals and procedures could be used, for example. Some examples are given to illustrate options open to a business.

Chapter 6

Several case studies are included here, described by the managers who implemented either BS 7750 or ISO 14001, detailing some of the trials and tribulations they had. Areas of the Standard that are open to different interpretations are also discussed, and the approach of both the client and the assessment body are included.

Again, the intention is to stimulate thoughts and ideas rather than to encourage the reader to copy their various approaches; these have worked well for one organization but may not be suitable for another.

Chapter 7

This chapter focuses on getting to know the auditor and how they function in the auditing role – the 'personality' behind all the questions – for he or she will ultimately decide whether or not your organization meets the requirements of the Standard. Myths and stories abound in this area of client-auditor relationship. The intention of this chapter is to open up this relationship to scrutiny and therefore allow the certification process to proceed smoothly via a co-operative approach, with auditors working alongside the client to add value to the system.

Chapter 8

Some organizations may well be undecided on which environmental system to implement and, ultimately, which certificate will be obtained. Within Europe there is a choice of EMAS, the Eco Management Audit Scheme, and ISO 14001, the International Standard. Should an organization aim for one or the other? Or indeed both? Can one be used as a stepping stone to the other? These are some of the questions addressed in this chapter; a book such as this cannot ignore the existence of such a significant, almost parallel, standard in the field of environmental management systems.

Appendix I

This appendix contains explanations of terminology used throughout the book.

Appendix II

Appendix II suggests titles for further reading. This book does not pretend to be comprehensive, covering all the environmental issues associated with the implementation of environmental management systems. For example, EMAS implementation and systems integration are not addressed in detail – further reading would be required. Additional reading is also suggested if the reader wishes to keep abreast of environmental management knowledge, as it is a very dynamic subject. Some addresses are given to assist.

Appendix III

ISO 14001 is just one standard in a series of environmental standards. This appendix lists the complete ISO 14000 series.

Appendix IV

This appendix contains information on the accreditation criteria for certification bodies. Chapter 4 explains the rules and regulations that certification bodies must operate to; these 'criteria' for operating as a certification body are mandated by accreditation bodies throughout Europe. Appendix IV lists the salient points of the accreditation criteria.

Acknowledgements

I am indebted to my employer, SGS Yarsley ICS Ltd, for bringing me into contact with Michael Forster of Butterworth-Heinemann, who suggested that there was a world-wide need for a book on ISO 14001 implementation. This coincided with some thoughts that I had, following an intensive period of delivering lectures, seminars and workshops to a wide spectrum of audiences, that some sort of guidance was required for those organizations that did not have the required environmental technical resources readily available: in particular, the so-called small to medium enterprises (SMEs).

That was in mid-1996, when there was a state of uncertainty in the world regarding environmental standards – ISO 14001 existed only as a draft standard. There was a need to produce a book with a certain amount of urgency because it was apparent that once the uncertainty had been removed, there would be a flood of organizations wishing to implement ISO 14001. This has certainly been the case and the market demand for environmental certification is growing rapidly.

I am also grateful to the many implementing organizations that I have worked with, both in the UK and overseas, as a third-party auditor, for valuable discussions on the interpretation of the Standard, and where possible I have included examples in the text.

Extracts taken directly from ISO 14001 are reproduced with kind permission of the British Standards Institute, and complete editions of this and other related standards can be obtained from them. Their address is included in Appendix III.

Several case studies of successfully certified organizations are included and I am particularly grateful for their contributions.

Other thanks go to my colleagues who gave me valuable advice during some difficult moments during the book's progress: especially Barry Holland from the Marketing Department, for some inspiration in the early days of writing the book; and my secretary, Allison Squire, for assisting in the word processing.

Finally, in my personal life, my thanks go to Christine, my wife, who put up with my hours of absence from the family circle and my teenage sons, Tim and Nick, for allowing me to use my personal computer for the writing of the book, when they had far more important games to play.

KW
November 1997

Chapter 1

Environmental management systems evolution

Introduction

The manager or director of a small to medium enterprise (SME), new to the concept of managing for the environment and environmental management standards, may well ask questions as to where such ideas and standards originate. Were theses ideas created by idle bureaucrats or were they derived from a perceived need throughout the industrial and commercial world? Were representatives from industry and commerce involved in the writing of such standards? Is there a market demand? What forces shaped the format, content and requirement of the standards?

This chapter will attempt to give answers to these questions. Initially, it will look back at the history of the growth of industrialization in the developed countries. It will then demonstrate that a combination of national and international forces at work, plus the legislative measures taken by successive governments, the rise of the stakeholder concept and green pressure groups, as well as the trigger of environmental disasters, have all played their part in the development of environmental management systems. Specifically, this chapter will describe how ISO 14001 has developed to become the International Standard for Environmental Management Systems that it is now.

National and international forces of change

Industrial Revolution – old and new environmental issues

If we look back to the Industrial Revolution, we might see this as the period when the inventiveness and innovation of human beings, and the resultant mechanization of manufacturing processes, began to have negative impacts upon the environment. Prior to this period, any negative environmental impacts tended to be localized – for example, forests were cut down without any future regard, but there was a limit to the amount of trees that could be felled manually by numerous foresters armed only with axes.

Immense changes to society began to occur and, consequently, vast amounts of non-renewable resources were consumed to support this industrialized society with no particular thought as to the longer term effects on the health of the population or the quality of the environment.

Successive governments brought in legislative measures to control the worst excesses of manufacturing pollution in the 19th Century. As the processes became more diversified and sophisticated in the 20th Century, ever more legislation became necessary to control these diverse industrial activities – especially as some of the health and environmental issues were of a potentially greater magnitude. Concurrent with this build up of legislation, the powers of the policing authorities grew – especially with regard to imposing greater financial penalties. It also became clear that the impact the Human Race was having upon the environment was also rising in magnitude – disasters occurred, lives were lost, and the environment was put at risk.

Gradually, forward-looking organizations, or even those that had been fined heavily following breaches of legislation, began to pick up the notion that managing for the environment was something that they would have to do if they wanted to stay in business. Such a provision began appearing at senior management meetings as an agenda item. In many ways, environmental issues began to be treated in the same way as commercial business decisions. As with all commercial decisions, there are two initial choices:

1   To do nothing now, but to keep informed of all changes in internal and external factors and review and modify the decision as necessary.

2   To be pro-active and take measures to prevent an incident: environmental planning management rather than environmental crisis management. This option is becoming increasingly preferable.

In the modern world, many of the Industrial Revolution's pollution problems have been solved. In highly-regulated, industrialized countries the sight of heavy industry belching out black smoke into the atmosphere and pumping unlimited toxic chemicals into rivers and streams is a thing of the past. For example, in London, the 'smog' of the early part of this century (caused in part by domestic burning of coal for heating as well as industrial pollution) will never return. However, there still are environmental impacts; it is only the technology that has changed. For example, the devices that we have invented to make our lives easier and more comfortable are causing their own particular environmental impacts – such as private motor cars, and the use of CFCs in air conditioners and refrigerators. Our lives have certainly been made easier by these devices but we are paying a high price for this convenience: increasing atmospheric pollution from cars and the threat to the ozone layer by use of CFCs.

Increasing legislation and financial penalties

A plethora of legislation now exists and is being continuously added to or amended. Keeping up to date with such national legislation is difficult; keeping up to date with international legislation is near impossible. For example, within Europe, over 320 environmental laws have been adopted by the European Union since 1967 (although some were only expansions or amendments of existing laws).

Principles such as 'making the polluter pay' were established in the courts, making the organization that created the pollution responsible for all the clean-up costs, including any consequential damages. As an example, consider the case of a holiday coastal resort suffering pollution, either from chemicals or oil. There will be loss of revenue to local businesses – due to the reduction in the number of tourists – and this will be treated as consequential damages. The costs of such consequential damage can be tremendous – far outweighing the actual clean-up costs of the original pollution.

Other financial penalties are also evolving:

- **Ethical lending:** The concept of only lending money to an organization that has demonstrated that it is environmentally responsible or, conversely, not lending money to an organization that has shown disregard for the environment.

- **Insurance risk:** Insurance companies are looking very hard at the every-day risks they are routinely underwriting. The environmental risks being run by a badly-managed chemical company, for example, can be daunting in the event of a disaster (with all the cleaning up and consequential costs, as referred to previously). Premiums are 'loaded' unless sound environmental management can be demonstrated.

Industry therefore sought a code of practice, a set of rules and a formula to grant immunity from bad publicity and prosecution – a standard to work to, to ensure sound management of environmental risks.

Major environmental initiatives

As a result of international initiatives on global environmental problems, a landmark conference – the United Nations Conference on Environment and Development (UNECD or the Earth Summit) – was held in June 1992 in Rio de Janeiro, Brazil, with representatives from some 150 nations attending.

Representatives recognized the need to involve all the major stakeholders in a sustainable future (see also later in this chapter). Scientists, research institutions, businesses, children, trade unions and local communities are all stakeholders in defining that future.

One of the outcomes of the Summit was Agenda 21 – an action plan setting out an environmental work programme into the 21st Century. In the UK, for example, local authorities have been given the challenge of co-ordinating that responsibility through local 'Agenda 21' activities (see Appendix I).

The globalization of trade

On a global scale, deep seated and accelerated change is now occurring. Nations and industries alike need to adopt a global approach to their economic policies and business strategies in order to remain competitive. Issues of both quality standardization and environmental responsibility are included in such strategies, as illustrated below:

Trans-national or multinational organizations
As an example, a corporate headquarters in Japan may want all its sites world-wide to achieve environmental certification. The organization may reason that the negative environmental activities on one site can impact on the image of the organization world-wide. There could be a weak link in the chain. Such an organization will have had management systems in place for many years to ensure that equal standards of production exist whether the plant is located in the UK, Taiwan or Mexico.

Movement of manufacturing centres from country to country
Industrialized countries – both in the West and the Far East – are undergoing processes of change and restructuring. Heavy industry in some countries is well into decline, especially over the last 10 years or so. The marketing, planning, designing and related 'intellect rich' activities can take place in the home country with all manufacturing taking place in locations never considered in the past. The reason they are considered now is costs – mainly low labour costs but also some economic assistance from the national government keen to encourage foreign investment and employment for its work force. Although the manufacturing processes are complex (for example, electronics or vehicle manufacturing) and the technology is sophisticated, a high degree of control can be exercised by using not only automation, but also personnel working to standardized procedures within management systems such as quality assurance (ISO 9000). Such systems can be 'transplanted' or 'transported ' to the new sites. This means that relatively unskilled labour can be used. Unskilled labour commands lower wages of course than the highly trained technicians in the corporate home country.

Of course, such quality assurance management systems have been developed and refined by the corporate headquarters in its native industrialized country.

Some years ago, it was said that only high-quality consumer goods could be manufactured in the Far East – Japan in particular. The decline of the car industry in the UK is a classic example of mass production being moved from traditional industrial centres to the Far East. It was believed that only the Japanese could produce the required high quality.

However, such preconceived ideas are now fading with the greater acceptance of the idea that the same potential for quality exists in every country – one need only install a culture of quality based on the framework of quality systems management: ISO 9000. These emerging industries in turn are looking to their suppliers to provide materials and sub-components of unimpeachable quality.

Since environmental considerations include reduction of waste (including less rejects and defects) then they look to their suppliers to reduce their environmental impacts. This in turn stabilizes prices as the supplier is reducing the costs of non-quality and its financial consequences. The supplier has more chance of staying in business and thereby guaranteeing long term stability of supply.

Associated with the above, a trans-national organization, negotiating with a foreign government to move a production facility, may well offer the existence of their ISO 14001 certification as proof that they are environmentally responsible. Again, the foreign government will be keen to demonstrate to its population and other world investors that it has included environmental considerations in its decision to allow foreign investments.

In conclusion, organizations may seek certification so that they can have a transportable management system (ISO 9000) for quality and also ISO 14001 for an environmental system. This will enable them to practise their business no matter where in the world and give confidence immediately that they are environmentally responsible and, therefore, more likely to be welcomed into a foreign country by that government.

The concept of the stakeholder

The 'stakeholder' is the concept of the 'customer' being not just one branch of human society but a whole spectrum of society who have an interest in the well being of a business – and do not want such a status damaged by poor environmental performance. Such a poor environmental performance may damage the image of the business, depress share prices, reduce market share, produce less profits and, perhaps, necessitate staff redundancies.

This spectrum of stakeholders includes:

- The parent company
- Customers

- The board of directors
- Environmental pressure groups

- The owners/shareholders
- Suppliers

- The investors
- The local community

- The employees
- Competitors

- The insurance company
- The general public

- The regulatory body

Thus there is a spectrum of 'customers' whose views and requirements need to be considered. It must never be forgotten that the general public demonstrate their environmental stance by exercising their power as purchasers. Such behaviour is based on the information available to them – which may not always be correct (for example, information received via rumours or bad press from competitors). If an organization wishes to be successful and profitable it must always take heed of this particular group of stakeholders and pro-actively manage its environmental impacts and demonstrate visibly that it is doing so.

Any organization should perform a simple stakeholder analysis – it can be illuminating – to demonstrate just who they are responsible to, and perhaps to ask what these stakeholders require of them!

Market encouragement measures

Market encouragement measures are measures designed to promote 'green' purchasing decisions on the part of the consumer. The main intention is to provide sufficient information for decision makers to be properly informed about the environmental performance of goods and services. ECO labelling and establishment of an environmental management system are just two examples of this approach and ISO standards are in preparation (ISO 14021 to ISO 14025 'Environmental Labels and Declarations' – see Appendix III). *ECO* is the abbreviated 'logo' derived from the practice of auditing organizations in the 1980s and 1990s for their ecological impact.

Disasters – as environmental triggers

In the Preface, the question was asked 'What is an environmental management system?' and a statement was made that all organizations have an impact upon the natural environment. Unfortunately, environmental impacts sometimes escalate into environmental disasters, which act as 'milestones' in the development of environmental awareness at the general public level and tougher legislation at the commercial level. These disasters have an increasing tendency to influence such awareness and legislative requirements on a global scale.

The human tragedy of such disasters is the most emotive of course and this is closely followed by considerations of the environmental issues. Examples of such disasters are:

1976 – Seveso. A chemical plant at Seveso in Italy accidentally released a quantity of dioxin into the atmosphere. Approximately 250 people showed signs of being affected by dioxin – including pregnant women.

1978 – English Channel. The supertanker Amoco Cadiz split into two, spilling crude oil into the English Channel, killing marine life and polluting miles of beaches on the French coastline.

1984 – Bhopal, India. At least 2,000 deaths were recorded from an accidental emission of methyl isocyanate gas. Approximately another 200,000 people were affected, with symptoms including blindness and liver failure.

1986 – Basle, Switzerland. A company using warehouses on the banks of the river Rhine suffered a large fire. Water used by the fire services to contain the blaze washed out pesticides containing mercury into the river. The effect on the marine life in the river was enormous. There were consequences for other users of the river waters (for example, potable water abstraction and amenity sites).

1989 – Alaska. The oil tanker Exxon Valdez ran aground and caused marine pollution and contamination of foreshores.

Disasters such as these are extreme examples of situations going drastically out of control – situations that could have been avoided using risk management. Huge financial and personnel consequences were involved for the organizations and the senior management.

These and other incidents resulted in strict environmental protection legislation being implemented across the world. The effect on public consciousness was profound and had a negative effect on the general public's perception of the chemical and manufacturing industries generally.

Incidents which involve major loss of life naturally and properly command huge media attention and prompt public enquiries. Such incidents are the trigger for new legislation aimed at preventing a repetition of the disaster.

Such incidents, which lead to tougher legislation, force other companies, unconnected to the offending original organization but nevertheless bound to comply by the new legislation, to take abatement measures quicker than they would have planned.

This leads to a reactive (rather than pro-active) style of management which can result in either 'quick fix' or 'end-of-pipe' technology, or taking measures which may not be in the organization's best interests over the longer term. Hasty decisions are the ones to avoid. In the worst case, businesses may close down entirely because the costs of the technology required in the short term to comply with the new legislation make it uneconomical for the business to survive.

Disasters, such as those above, make the headlines in the international media. However, at a lower level of publicity, many organizations and individuals are concerned about the issue of sustainability.

Sustainability

The general public is becoming increasingly aware of the issue of sustainability. The very science and technology that has been developed to give us the lifestyle that we enjoy now, especially in the Western industrial world, has left us in a position to hand down an environmental legacy to our children and grandchildren. This legacy can either be a better or a worse lifestyle – the choice is ours!

To date we are giving our children the prospect of:

• Nuclear contamination – either accidents or potential nuclear warfare

• Global warming – accumulation of greenhouse gases

• Ozone depletion – destruction of the Earth's defences

• Deforestation – destruction of the tropical rain forests

- Acid rain – acidic gases such as sulphur dioxide and nitrous oxides from the burning of fossil fuels dissolving in water to form acids

- Depletion of resources – especially fossil fuel reserves and minerals

The Earth does not have infinite resources. Indeed, we have been likened to a spaceship – 'Spaceship Earth' – travelling through the universe, and this is a good analogy, as a spaceship has only finite amounts of fuel, oxygen, food and water to support the astronauts on board throughout their journey. The Earth is no different. We need ever increasing amounts of resources to supply an ever increasing population with an appetite for scarcer resources on our somewhat repetitive journey around the sun. Our only apparently limitless source of energy – heat and light – is from the sun. However, the prospects of fully harnessing such an infinite resource is some time off, or the stuff of science fiction, depending upon the individual's point of view! All concerned people believe that we should act now to ensure the sustainability of the future.

The issue of sustainability is too broad a subject to treat in this book, but a summary of issues of sustainability is given below.

Population
In Western society, although population growth is slow (perhaps 6%) the pattern of population is changing. More and more people are either staying single or living in smaller groups or separately. This puts pressure on demands for more housing stock, land, energy and goods consumption.

Global atmosphere
There are two main concerns :

1   That emissions of carbon dioxide and other greenhouse gases, such as methane, may result in warming the earth's surface, leading to changes in climate and a rise in sea level.

2   That the production of CFCs (Chlorofluorocarbons) will deplete the ozone layer surrounding Earth and let the sun's harmful ultraviolet rays penetrate through.

Air quality
Pollution from vehicles and industry affects urban air quality. Emissions from road transport have doubled over the last 20 years. Acid rain caused by sulphur emissions contributes towards the acidification of freshwaters and soils.

Freshwater

Abstraction of surface and groundwater places a demand on our water resources. Even in a temperate, wet climate as in the UK, periods of drought are not unknown. This is due to several factors, including increased consumer and industry demand.

Groundwater quality is also a serious issue because of the difficulties of cleaning water once it has been polluted.

The sea

Marine issues include: water quality, especially around estuaries in industrial areas; dumping of waste at sea; over-fishing (which has led to dwindling fish stocks, with some species thought to be at unsustainable levels); over-extraction of marine aggregates; and environmental impacts associated with oil and gas exploration.

Soil

The main concerns relating to soil are: irreversible erosion; loss of organic matter; increase in acidity and the effect on soil quality (which is affected by air pollution); contamination of land; and irresponsible agricultural practices.

Land use

Land resources are finite – especially in countries with a small land mass and a large population. It requires skilful and careful planning to maximize the use of land, particularly in those countries with an increasing demand for out-of-town retail 'malls' which tend to use virgin 'greenfield' sites rather than inner city, 'brownfield' sites.

Minerals – including fossil fuels

Demand for minerals and aggregates is expected to continue to increase. Some of the demand can be met through recycling but more efficient utilization is the real answer.

Wildlife and habitats

Due to deforestation and use of greenfield sites, many more species of flora and fauna have been made extinct over the last 50 years than in our entire history.

Waste

However well managed, waste always has the potential to cause harm to the environment. The acceptance of the existing 'end-of-pipe' hierarchy of waste management could be construed as an acceptance of the unsustainable patterns of production and consumption.

Therefore, priority must be given to reducing the amount of waste produced and to adopting better waste-management practices. There is a hierarchy of waste-management options:

First   Reduce overall consumption of resources.

Second   Consume selectively. Aim for maximum possible use and consider aspects of durability, repair and recycling. Purchase second-hand, lease, hire or share wherever possible.

Third   Minimize the generation of waste.

Fourth   Re-use whenever appropriate.

Fifth   Recycle. This includes composting for biodegradable materials.

Sixth   Recover energy. Use heat exchangers, for example.

Seventh   Dispose to landfill sites only as a last resort.

To ensure that waste-management practices become more sustainable, the emphasis is being placed on the options at the top of the hierarchy.

(The wider issues of 'true' sustainability are beyond the scope of this book and, therefore, no advice is given to an organization striving to achieve this goal. However, it can be said that an environmental management system can contribute to sustainability by focusing an organization's resources onto reducing impacts on the environment.)

BS 7750 and ISO 14001

BS7750 was a British Standard designed for environmental management control. Other similar national standards were also written (in Spain and Ireland, for example). A 'consumer' demand for an internationally recognized standard led to the development of ISO 14001, which was firmly based on the requirements of BS 7750.

The birth of BS 7750

It was stated in the previous sections that the combination of increasing legislation, punishments through the medium of fines or imprisonment, stakeholder pressures, the effects of disasters and the usefulness of transportable environmental management systems, all contributed to industry searching for a set of rules and guidance benchmarks that it could work to. If adhered to, such measures would prevent an incident or a disaster and guarantee escape from prosecution. One of the consequences of these concerns and pressures was requests to standard-writing bodies to produce a standard for managing the environmental effects of an organization. As a result, the British Standards Institute (BSI), as a world-respected standards body, in conjunction with many other committees and interested parties, developed and produced BS 7750: 1992. This was the world's first environmental standard. The purpose? To get businesses out of a fire-fighting mode and to make them pro-active in addressing environmental, legal, image and financial risks.

Although such bodies had been involved with the standardization of quality management issues for many years (ISO 9000), designing and writing a standard for environmental management was a very different issue.

Many such bodies and (for example in Europe) regional bodies responded by establishing advisory groups and technology committees to enquire into the market need and then to begin the standard-writing process.

This is how the BSI in the UK came to create its technical committee and how BS 7750:1992, the world's first environmental management system, came into existence.

BS 7750 was produced by a committee representing UK industry, professional and learned bodies, government trade unions, consumers and voluntary bodies. It was written as a specification – a requirement-based document designed to be assessed – with a certificate to be awarded to users meeting its requirements. In the UK, the BSI published BS 7750:1992 and a pilot programme with some 200 implementing companies was set in motion. The pilot programme raised many issues:

- Clarification of the role of occupational health and safety issues within an environmental management system

- The meaning of the word 'significance' in relation to environmental impacts

- Levels of performance expected of an implementing organization

- Competence and expertise of certification bodies and individual auditors

Based upon the feedback from these organizations, the Standard was modified and re-published in 1994.

During 1995, certification bodies were invited to work with the United Kingdom Accreditation Service (UKAS), 'witnessing' assessments for accreditation purposes. Up to this point, all certification was unaccredited, and UKAS was pioneering accreditation (see Appendix I). The witnessing, or auditing, exercise was a team approach between UKAS and the certification body, allowing the identification of the competencies necessary for a certification body to operate in this new area of certification. As well as witnessing assessments, UKAS also examined the certification body's head office functions. A focal point of the audit was a test of the integrity of the procedures which ensured that only properly-qualified, experienced auditors were assigned to audit individual organizations to ISO 14001.

During this period, EMAS (Eco Management Audit Scheme) was also developed and is discussed separately in Chapter 8.

International dimensions – the birth of ISO 14001

World-wide demand for accredited certification to an international standard was growing. In 1991, ISO (the International Standards Organization) received the same pressure as national and regional bodies from its members and established a Strategic Advisory Group on Environment (SAGE) to investigate all areas of environmental management and performance. SAGE recommended that a new technical committee be established to develop international standards in environmental management.

Thus TC (Technical Committee) 207 was established in 1993 to develop and produce the standards proposed by SAGE and to investigate the possibility of developing more related and supporting standards. TC 207 created six subcommittees and one working group to work on different areas of environmental management.

It was agreed that TC 207 should have a formal link with other TCs (such as TC 176 – Quality Management and Quality Assurance) because there were strong linkages between such management standards. The scope of TC 207 was agreed as covering all standardization in the field of environmental management tools and systems. Test methods for pollutants and effluents, environmental performance levels, and standardization of products were excluded. (See Appendix III for a list of these standards.) TC 207 was composed of representatives from business, industry, learned bodies, consumers and many other groups. The committe was mindful that ISO 14001 had to be written in generic terms, and considered:

- Applicability to manufacturing and service organizations alike.

- How to fulfil the needs of small and medium sized enterprises. (*ISO 14002 Management Systems – guidelines on special considerations affecting small and medium enterprises* was an outcome of this consideration.)

- The need to avoid trade barriers as well as the different approaches to legal requirements and their enforcement throughout the world.

In response to consumer demands for ISO 14001 certification, the decision was taken within the European accreditation bodies, so as not to slow the progress of international environmental certification, that organizations could be assessed by certification bodies against the draft standard DIS/ISO 14001 during 1995. In the UK for example, UKAS authorized such certification from December 1995.

Certification to the draft standard was 'seamlessly' converted to ISO 14001:1996 at the next surveillance visit by the certification body, following full publication of the Standard. 'Seamless' transition means that the client should not be aware of the changes in terms of extra costs or extra onerous requirements. The changes from the draft to the published standard were textual rather than of technical content, thereby allowing such conversion.

In Europe, ISO 14001 was published as a European Standard BS EN ISO 14001: 1996 by CEN (the European standards body). By rules established between national European and international standards bodies, once a standard is accepted as a CEN standard then the equivalent national standards must be withdrawn within a six-month period. Therefore, BS 7750, which is a British national standard, should have been withdrawn by 31 March 1997. However, recent developments within the European Commission resulted in an extension of the date of withdrawal of BS 7750 (and other national standards) until 30 September 1997. The reason for this is that, as yet, the EC has yet to recognize EN ISO 14001 as being appropriate to fulfil EMAS requirements. This is addressed further in Chapter 8. (See Figure 1.1.)

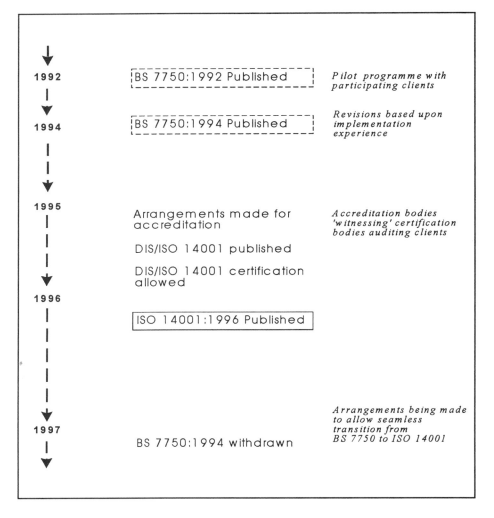

Figure 1.1 Key dates and events in development of ISO 14001

Differences between BS 7750 and ISO 14001

Detailed comparison of ISO 14001 with BS 7750:1994 showed differences but the intention, and the spirit, was identical. Indeed in the changeover period for those organizations with BS 7750 wishing to change to ISO 14001, the transition was relatively straightforward: documentation changes and some changes of emphasis in the wording of manuals, procedures and operational practices were all that was necessary. (In the UK, UKAS produced a briefing note with a set of 18 points needed to be addressed in such a transition – see Chapter 4 and also Appendix II.)

Reasons for seeking ISO 14001 certification

Previous sections have addressed the broader issues of why environmental thinking is now higher on the agenda for an organization doing business in an increasingly competitive world, and how in response to this, national standards such as BS 7750 became refined into an international standard to assist organizations in meeting the requirements of stakeholders in a world demanding ever higher standards of environmental responsibility. Therefore, asking an organization in the 1990s why it chose to become certified to ISO 14001 will probably prompt the following replies:

• To gain or retain market share via a green corporate image

• To attract more ethical investment

• To reduce insurance risks

• To reduce prosecution risks

• To reduce costs

The reasons may not necessarily be in this order of importance. However, the fact is that cost savings tend to be low on the list of responses. Image and potential loss of business are cited the most frequently. Cost savings tend to be overlooked, yet this is an area where implementing organizations have unplanned benefits. These reasons will be considered in turn and discussed in detail below.

Market share and corporate image

A 'green' image is an important marketing factor, and even in a situation where a positive image is not particularly high on the agenda, an image of being 'non-green' is a definite marketing disadvantage. In any marketing opportunity, this concept should be kept in mind, especially with the younger generation, who have been schooled in environmental issues. Today's young people are tomorrow's consumers, with significant influence on where money is spent. Some examples from around the world, where having a 'green image' is of economic importance, are given below. Some organizations have seen their image tarnished in the eyes of the public – usually the end user and purchaser of their products or services. Such a tarnished reputation has led to a drop in the market share.

• The **Asia Pacific Institute of Environment Assessment (APIEA)** experienced a sharp rise in the level of interest being shown by a broad range of industrial sectors in environmental management systems – driven, it was supposed, by such south-east Asian companies losing overseas contracts through failure to meet increasingly stringent national environmental standards in the USA and Europe. Thus the fear that failure to comply with international standards, which are anticipated to be widely adopted in the West, could have sobering economic consequences provoked action.

• In **Indonesia**, the Government is actively assisting companies there to perform environmental audits and making such audits mandatory for those that have a pollution incident.

• In **Malaysia**, the government announced its intention to make it compulsory for all companies to institute self-assessments of their environmental performance and to have the assessments verified by an external body (in effect requiring certification to ISO 14001).

• In the **UK**, a home improvement and garden centre retailer was held accountable by green pressure groups, including Friends of the Earth, for irresponsible sourcing of timber. Although not directly involved in felling the trees, the retailer was seen to be a prominent and accessible target for public concerns, and was being held accountable for the environmental consequences of its purchasing. Thus, this retailer was forced to consider what responsibilities it had towards any of its products and what environmental impact such products had at any stage in their life cycle – in the acquisition of the raw materials, in the manufacture of the product, in its use, or in its disposal.

- The **Forestry Stewardship Council** was set up to promote active encouragement of forestry operations to achieve their own forestry certification – but this certificate has many commonalities with ISO 14001 and dual certification is currently occurring in Scandinavia.

- One **multinational oil company** lost much revenue when consumers refused to purchase petrol at filling stations, following adverse publicity over the company's decision to dispose of a redundant oil rig. Interestingly, above and beyond the decision for disposal, was the lack of consultation and the company's high-handed treatment of its stakeholders. This, additionally, sent out a message of low priority for environmental issues for the company, and was probably more significant in influencing consumers' negative purchasing decisions.

Ethical investment

Increasingly, investors are asking questions about the environmental performance of an organization prior to any investment. Indeed, *ethical* or 'green' investment is an evolving aspect of investment management. It means only investing in companies that meet certain ethical criteria (in terms of their sphere of operation or their corporate social policies) and which implement good environmental practices or provide products or services which contribute to environmental improvement. Ethical funds initially arose in response to church-based investors' wishes to avoid companies involved in a range of activities (such as the manufacture of armaments or tobacco products) and then spread, in the 1980s, into environmental areas as public awareness of environmental issues heightened.

Many funding organizations (such as insurance companies and pension funds) now have separate 'green' investment funds for those investors who are prepared to invest this way (rather than for those investors who only want a good return on capital and don't care where their money is invested). Investment bodies, or trusts, must have a mechanism to screen and audit companies so that they can reassure their investors that such companies exercise environmental probity.

A typical screening involves a blend of negative and positive approaches to green investment, as desribed below.

Negative approach

The negative approach is applied to six specific activities:

1   The alcoholic drinks trade

2   The armaments trade

3   Gambling

4   Nuclear power generation

5   Publication of pornographic material

6   The tobacco trade

Typically, those companies who obtain in excess of between 1% and 10% of turnover from the above industries will fail the negative screen and will not be included in the investment portfolio.

Positive approach

Companies that have passed the first screening are then subjected to a more positive approach where the company is considered as a whole and its complete performance is evaluated. Because some evaluation of the organization's commitment is required, such an approach requires active dialogue achieved by visits to the company and face-to-face meetings with all levels of company employees.

Both 'process orientated' and 'products and services orientated' criteria are evaluated during the investigation. Both categories help to identify how companies are working to reduce environmental damage. Such damage typically arises from the following:

Process orientated criteria

¤   Emissions and effluents

¤   Waste

¤   Energy consumption

¤   Raw materials

¤   Transportation

¤   Land use

Products and services orientated criteria

(The objective of this aspect of the assessment process is to identify companies that offer products or services that have a beneficial impact upon the environment.)

¤   Emissions

¤   Controls

¤   Recycled products

¤   Renewable energy

¤   Environmental advice and consultancy

Determining a product's purpose is generally simpler than establishing how it is made – so this research aspect of the assessment process is more reliant on published information than an investigation of the manufacturing processes. Such research covers:

- **Environmental impacts**: The extent to which products and services promote environmental improvement by reducing harmful emissions, waste, or resource depletion.

- **Research and development**: The commitment to the research and development of products which may assist environmental performance in the future.

- **Packaging and labelling**: Whether packaging is minimized; whether packaging is biodegradable or reclaimable; what information is given to the consumer about content.

Additionally, a corporate environmental policy is looked for, as well as a recognized environmental management system. The existence of such a system is important during the positive approach audit.

In conclusion, this style of ethical screening has been in use for some years now and does have a steady growth profile. There is a high demand from individual investors as well as corporate bodies – banks, pension funds, building societies – to only invest in such companies. The only barrier to further growth appears to be the low level of publicity about the existence of ethical funds, and the consequent low level of awareness amongst private investors.

Insurance risks

Increasingly, insurance companies are reviewing the structure of how they set the levels of their premiums. An environmental impact assessment will certainly be used in determining the exposure they will be open to, and the amount of premium they will charge to cover the risk. In November 1995, 54 international insurance companies signed up to the UN's *'Initiative on Sustainable Development'* and the insurance industry pledged to 'incorporate environmental considerations into their risk management'. As a result of losses faced by the insurance industry in the US throughout the 1980s, insurers have been increasingly reluctant to insure some classes of environmental risks – particularly gradual pollution.

Lloyds of London suffered heavily – for the first time in its history – during the 1980s and early 1990s. This was partly due to natural disasters such as: hurricanes Hugo (1989) and Andrew (1992); windstorms across America and Europe; and typhoons in Japan but mainly due to pollution incidents (such as asbestos-related claims and the Exxon Valdez disaster). This combination nearly overwhelmed Lloyds' ability to cope.

The perception of insurers is that the greater the technical complexity of an undertaking or business venture, the greater the danger and risks to the environment. They focus on consequences rather than probabilities. Thus, the onus is upon industry to educate all interested bodies in order to improve understanding of the fact that environmental management systems can control technical complexity to reduce probabilities. Certainly, lack of standardization in environmental reporting is one hurdle to greater understanding.

One idea also mooted would be to have environmental management included as one element of an organization's assessment for ranking on stock market listings. A good start to this process would be for businesses to report consistently for liabilities – a process which has begun in the USA – and a push for a more structured reporting style through an Environmental Management System such as ISO 14001.

Prosecution risks

A pro-active approach to environmental management can reduce the risk of fines or imprisonment by demonstrating that the organization exercises due care and diligence.

The 'polluter pays' principle, previously mentioned briefly, was introduced by the EC Environment Programme in 1972 and was adopted as European Law in the Single European Act 1987. It states that the polluter, or producer of the environmental damage, must meet the financial costs of the pollution caused. The principle has been upheld in court cases. Evidence is gathering to show that having an environmental management system can be used in mitigation, resulting in reduction of costs.

In the USA, the US Department of Justice has stated that the existence of an environmental compliance programme will be taken into consideration when deciding whether to prosecute a company for environmental violations. The US Environmental Protection Agency (EPA) is currently deciding on the possible use of ISO 14001 within their legal framework. Whilst it is unlikely ISO 14001 will be a mandatory requirement, it appears the EPA are considering offering reductions in penalties for breaches of consents if organizations hold ISO 14001 certification.

In the UK, inspectors from the Environment Agency have the powers to pay a compliance visit to an organization's site at any time. Having a well managed environmental management system, however, should enable an organization to demonstrate sound environmental practice. If the inspector does find something amiss during a visit, having a system in operation will be looked upon more favourably than having no system at all. This may be reflected in the levels of fines imposed.

Cost savings

Any organization investing in plant, machinery, personnel and so on, expects a return on its investment (in the form of cost savings) as soon as possible. It is no different with implementing a management system and, initially, although ISO 14001 is sought for a variety of reasons, a return on the considerable investment is also looked for. Data gathered to date shows that there are savings mainly due to management being more focused in its activities when looking for means to reduce costs – predominately energy and raw materials.

In the UK, for example, a landfill tax was introduced in late 1996 which effectively put up the costs of disposal of inert and special wastes to landfill. The objectives of the new tax are:

- To ensure that landfill waste disposal is properly priced. Waste disposal companies are then in a position to pass the additional costs on to waste producers. This in turn means that waste producers will be made aware of the true costs of their activities and so have an incentive to reduce waste and make better use of the waste which they produce.

- To apply the 'polluter pays' principle and promote a more sustainable approach to waste management, in which we produce less waste and re-use or recover value from more waste.

This was a demonstration of the UK Government's commitment to extending the use of economic instruments as a means of achieving environmental objectives in a cost effective manner. This tax – and there will surely be others to come – will force organizations to rethink their waste management strategies. One effect is that inert waste (such as construction and demolition waste) can be landfilled at a quarter of the cost of other waste. *Inert waste* causes less environmental pollution to the extent that it does not generate methane when it is landfilled. However, if inert waste is contaminated by a certain percentage of special waste then the whole amount is classed as 'special' and the higher charge applies to all of the waste.

This must focus any organization's attentions on how it segregates its waste as well as the quantity of waste it is producing in the first place (arguably the area where it should improve first). Within Europe there is a plethora of legislation. One particular directive ('*The Packaging Directive 94/62 EEC on Packaging and Packaging Waste*') was adopted by the European Parliament and Council on 20 December 1994 and member states had to introduce implementing legislation by 30 June 1996. The Packaging Directive has mandatory requirements called 'essential requirements'. Within these essential requirements are statements to say that packaging is to be designed and manufactured in such a way as to facilitate its re-use or recovery and to minimize its environmental impact.

Thus, as an example, in the UK most businesses that handle more than 50 tonnes of packaging annually will have a legal obligation under the Producer Responsibility Obligations (Packaging Waste Regulations), subject to turnover, to prove that a proportionate amount of packaging waste has been recovered and recycled. Organizations must establish return, collection and recovery systems for packaging. Collection of materials will not be enough. It will be necessary to prove that specified tonnages of packaging waste have been recovered and recycled. This includes reprocessing into new material, composting and incineration with energy recovery.

Organizations, unless they opt for a collective scheme, will have to supply data on their waste material to the appropriate regulatory environmental body. (In the UK this is the Environment Agency.) This data can be openly inspected by the public. Some organizations will not be at ease with this disclosure. Commercial organizations are now in business to assist companies discharge their obligation through a collective scheme that consolidates individual organizations' waste data. Such schemes demand a fee, of course, for the service provided.

Thus, somewhere in the packaging chain, costs can be reduced by better design of packaging and better-managed ways of using packaging – exactly what an Environmental Management System can achieve.

Summary

In conclusion, ISO 14001 had its origins in BS 7750 (the world's first national environmental standard). It is now the International Standard for environmental management control, in a world where environmental issues, such as sustainablility, are gaining momentum. Many organizations are putting environmental risk assessment high on their agenda when considering risk management in general. Environmental issues are integrated within their overall business strategy and the views of a whole spectrum of stakeholders are considered during the strategic decision-making process.

There are powerful driving forces behind environmental management. The need to demonstrate to customers environmental responsibility – to have a 'green' image – and to reduce prosecution risks are perhaps two of the most powerful. Barriers to implementation by smaller organizations exist, even for those with quality assurance management standards in place because, although there are many similarities in the two systems, there are some fundamental differences in philosophy which must not be overlooked. Even in the short life span of BS 7750 and now the early days of ISO 14001, there is a level of misunderstanding of what the Standard is for and what it will achieve for an organization. This is illustrated by the following list of questions which are often, and repeatedly, asked of auditors by potential clients:

Q1   What can a company do when it has addressed all its significant impacts, has shown improvements and has met all its objectives and targets?

Ans.   *This is the real world and no organization ever achieves such a state. Also, don't forget that once the first significant impacts have been addressed, the other less significant impacts become the significant impacts and a new cycle begins.*

Q2   How does an auditor assess two identical organizations who make identical products yet have different objectives and targets? (One is producing more atmospheric discharges than the other, for example.)

Ans.   *This is the real world and no two organizations would ever be in this idealistic scenario. Even if they are operating the same process, there are so many variables in a complex organization that this situation would never arise.*

Q3   Does certification indicate that a company's products/services are environmentally superior?

Ans.   *No. That was never the intent of the Standard. However, certification indicates that, during the manufacturing process, environmental issues were considered and may well have led to an improvement in that product (due to the focus on its environmental impact at the design stage).*

The next chapter will address these and other questions.

Chapter 2

Concepts and the 'spirit' of ISO 14001

Introduction

Chapter 1 addressed the gestation and birth of environmental management systems – the triggers of local, national and global issues – and described how, in particular, ISO 14001 developed into the international standard that it now is.

Before describing the steps an organization needs to take to implement the Standard itself, some of the underlying concepts are considered so that an understanding may be reached as to why the clauses of the Standard are written as they are. An understanding of the intention of the Standard – the 'spirit' – is also considered. For, unless the requirements of the Standard are understood at an early stage, the resultant environmental management system may have weak foundations. Such a system will not give the performance improvements intended – thus wasting the resources of the implementing organization.

The structure and the purpose of the clauses and sub-clauses are addressed in straightforward language and, where possible, simple, illustrative examples are given. Chapter 2 sets out the framework whereas Chapter 3 details all the steps necessary for practical implementation of ISO 14001. The reader who is unfamiliar with environmental management systems should therefore read Chapter 2 first so that the detailed approach of Chapter 3 can be better understood.

The first parts of this chapter explore the concepts behind what any environmental management system should set out to achieve. Later, attention is focused on how such concepts are refined for ISO 14001 environmental management systems: the reasons for the clauses; why they are phrased in the way that they are; and what they require of an organization in practical terms.

Concepts of environmental control

On a personal level

We all have an impact on the environment by the mere act of living from day-to-day. An environmental management system, in its simplest form, asks us to control our activities so that any environmental impacts are minimized. This broad and simplistic approach has its merits. However, such a loose, unstructured approach may lead us to improve in the wrong direction or, indeed, may leave us without any clear direction at all. On a personal level, it is tempting to control and minimize those impacts we feel we can tackle easily. Perhaps our attitude towards environmental issues is influenced by a topical environmental event reported in the media, and therefore, we can be influenced to act without thoroughly understanding some of the more complex issues.

Thus, as individuals, we may focus on, and minimize, environmental impacts which are trivial in nature compared with other impacts (which are far more significant and require more considered thought processes). As an example, we may commit ourselves to a futile exercise without attacking the root causes of pollution or the use of non-renewable energy sources. We may, in our working environment, always: re-use paper (writing on both sides); re-use paper clips; recycle plastic drinks cups. Such measures require only a little thought, basic discipline and very little personal physical effort. Yet we may use our car to drive to the office in a city – contributing to air pollution, traffic congestion and so on – when alternative transport could be used (for example, the humble bicycle or public transport). However, this latter option for environmental control requires much forethought (such as planning the journey time around bus timetables). There is some personal inconvenience and physical effort in this choice as well as some loss of freedom and flexibility. This is not to say that re-use of paper clips should be discarded as an environmentally responsible option but that we must be aware of its environmental significance compared to the other, more significant, environmental impacts.

At a business level

Moving from individual actions to corporate actions, and using the analogy above, unless a structured approach is taken the organization may focus on what it believes to be its environmental impacts, a belief based upon 'gut feel' and ease of implementation. In reality, this does not address real issues but promotes a 'green' feel-good factor or perceived enhancement of image – both internal and external to the organization – which is not justified. For example a company engaged in the extraction of raw materials by mining may have an environmental objective to save energy. By implementing a 'save energy by switching off lights' campaign in its site offices it may feel it has achieved 'green' status and may proudly boast of such an environment-friendly approach.

Indeed, there will be some energy saved by administration personnel (being of a disciplined state of mind) switching off lights and heating when they are not being used for long periods. However, such savings in energy are trivial compared to the massive impact that the mining industry has on the environment: the visible impact of the site and surrounding land; the associated increased noise levels from the operation of such a site; the high use of energy both in extraction technology and transport activities; the use of chemicals in the purification process; and of course, the use of non-renewable resources (the raw material that is being mined). Unless the mining company considers the relative scale and significance of environmental impacts, then by claiming to be 'green' it has really missed the whole point of environmental control and impact minimization.

Thus, this concept of *significance* is fundamental and must be at the heart of any environmental management system. An organization must move away from this 'gut feel' approach to a structured system that demands as a minimum for the organization an understanding of the concepts behind and strong linkages between:

• Identifying all environmental aspects of the organization's activities

• Using a logical, objective (rather than subjective) methodology to rank such aspects into order of significant impact upon the environment

• Focusing the management system to seek to improve upon and minimize such significant environmental impacts

It should be noted that the criteria used for attributing significance to environmental impacts should be clearly defined. The process of evaluating each aspect against the criteria should be readily apparent. This methodology of *ranking* of significance is very important – it must be robust and withstand scrutiny. Ranking is examined in greater detail in Chapter 3.

For example, for one company the most significant environmental impact could be the sending of mixed waste to a landfill site – a fairly common environmental aspect shared by most manufacturing organizations. The organization must then decide on an objective to aim for in reduction of such waste. This objective could be to reduce progressively waste sent to landfill by 3% per annum. Individual targets to support this objective could be set to: progressively segregate waste; recycle a certain percentage; and, perhaps, sell off a certain percentage of segregated waste (for example, brown cardboard). All these measures would reduce the number of skips of waste sent to the landfill site.

The reader will note that a figure of 3% per annum was chosen at random by the author. The organization itself should use more than just a random figure. The actual figure must be based upon what is practicable in the situation and what other similar industries are achieving. The organization should certainly use easy-to-measure data to support this waste minimization objective and this is further examined, developed and discussed in Chapter 3.

Remembering that any environmental management system is seeking to place controls upon its environmental impacts, then it is only common sense to have a plan for monitoring and measurement of controlling activities. Such a plan should readily show any deviations from the targets during a review, so that if a problem does occur, then the appropriate remedial or corrective actions can easily be taken.

This is environmental control. An environmental management system such as ISO 14001 merely provides the framework to allow such controls to be exercised in a structured and controlled way. By documenting such a system, personnel operating it have a framework to: work around; hang ideas onto; follow what is documented; record what was done; and learn from any mistakes that were made.

The spirit of ISO 14001

In the simplest of terms, and condensing the whole concept of ISO 14001 into one sentence, we can say that fundamentally the Standard requires an organization to:

> *Control and reduce its impact on the environment.*

No more, no less. Going further, again in simple terms, the Standard requires an organization to state how it goes about controlling and reducing its impact on the environment: doing in practice what it has stated; recording what has occurred; and learning from experience.

What obligation does this impose upon an organization? ISO 14001 requires an organization to control its impacts on the environment. All aspects of business activity cause changes in the environment to a greater or lesser extent. Organizations deplete energy sources and raw materials and generate products and waste materials. These changes are referred to as *environmental impacts*. ISO 14001 defines an environmental impact as:

> *Any change to the environment, whether adverse or beneficial, wholly or partially resulting from an organization's activities, products or services.*

Identifying and assessing the significance of environmental impacts is a critical stage in an organization's preparatory stages for ISO 14001. Thus the organization needs to understand that by operating its processes, by manufacturing its products or supplying its services, it is depleting natural resources and using non-renewable energy sources. At the same time it is also producing by-products in the form of waste materials.

This should not however promote guilty feelings within the organization! The Standard does not require organizations to feel guilty and apologetic. There is no hidden agenda to close the business down. The Standard requires management, by forethought and action, to use less scarce resources by better planning, use recycled materials and perhaps operate the process differently. By achieving the requirements of ISO 14001, an organization demonstrates it has a management system that will control and reduce its environmental impacts over a period of time.

An element of the controls required by the Standard will be dictated by the demands of legislation. Thus, to keep within the law, the organization will wish to ensure that all regulatory and legislative requirements concerning its environmental performance are satisfied. Increasingly, however, organizations are seeking to go beyond those legal requirements in order to ensure that their environmental integrity (of activities, products and services) meets the expectations of customers, consumers, investors, employees, environmental interest groups and the public. So, in effect, compliance with the law is mandated by the legal authorities. Controlling environmental impacts is also mandated – not by the legal authorities but by the stakeholders – as there is an inherent requirement, from the above discussion, to improve or minimize environmental impacts.

An interesting question is often raised by those personnel planning the early steps of ISO 14001 implementation. The question is along the lines of imagining the scenario when there is a point in time when all the environmental objectives of the organization have been fulfilled and where, perhaps, further improvements would be subject to the law of diminishing returns (remember, the Standard does not aim to bankrupt or close down business). What does the organization do next? Will ISO 14001 certification be lost? Does the organization attempt to improve in environmentally trivial areas, performing a meaningless paperwork exercise merely to generate evidence that the system is still alive, in order to retain certification?

The reality is that once the initial significant environmental impacts have been controlled and minimized, the other hitherto less significant impacts become more significant and a new cycle of improvement begins. Thus the cycle is never-ending and there is continuous improvement of the organization's environmental performance. The environmental management system is doing the job for which it was designed and implemented.

Two illustrations from history demonstrate (with hindsight) that our knowledge of environmental issues is usually flawed and that we, as individuals and organizations, acted in an environmentally responsible way based upon the knowledge available to us at that time:

- The use of CFCs (ozone-depleting chemicals) was not thought to be an environmental issue. We now know that it has a highly significant global environmental impact with possible long-term damage to our quality of life on Earth.

- The use of asbestos (especially blue) was at one time not thought to be an environmental issue nor a safety hazard. Asbestos is a material with unique properties and was widely used in insulation and construction. The environmental impact of removal and safe disposal of asbestos is often a highly significant impact due to such widespread use.

Thus, the rules can change. New knowledge comes to light and new, tougher legislation will always be around the corner. Therefore, this status of 'zero significant environmental impacts' will never occur.

Another query, often put forward in the early stages of implementation, concerns two hypothetical organizations – although they manufacture identical products one has a higher impact upon the environment than the other:

- Producing more waste to landfill

- Using more energy due to older plant

- Has more breaches of legislation – violations of discharge consents, for example

- Is visually offensive due to old, badly-sited buildings

- Has more smell and noise nuisance

How can both organizations achieve certification if one is clearly not as environmentally responsible as the other? Unfortunately, the answer is that they are both equally environmentally responsible if they are certified to ISO 14001. They are both equally committed to environmental improvement but the starting point for this environmental improvement is different for each of them.

They will both have the same potential environmental impacts but the landfill and energy-usage question may be due to better or worse technology – one organization may have access to capital from a parent company and will therefore perform better in these respects than their poorer competitor.

The environmental improvement objectives of the 'poorer' company may, in fact, be similar to their more affluent competitor but, for example, percentage improvement figures may be of a lower order. One longer-term objective could be to match the environmental performance prevalent within the organization's own industrial sector. This objective is very much dependent upon an organization's economic performance.

An environmental management system does not seek to be comparable – it proves only that each organization is seen to be committed to taking appropriate and practical steps to reduce their environmental impacts (within their individual capability and level of technology).

Providing that both organizations can demonstrate such commitment, the certification body will allow certification. This is the concept of the environmental management system: it is an improvement process, rather than a method for stating that, at any one point in time, one organization is performing better than another. (It should also be pointed out, of course, that the idea of two such identical organizations is fiction and could not form a real-life situation!)

The clauses of the Standard evolve from this simple common sense investigation of an organization's activities with some additions and enhancements (for example, ensuring mechanisms are in place to make a company aware of new and impending legislation). As the reader will note from reading the Standard itself, it is not a long document and is written concisely, with only six main clauses. It is generic in style, as it is intended to be applicable to any manufacturing or service industry.

The following section looks at the Standard in more detail.

Clauses of ISO 14001 – the intended purpose

Having looked at the concepts of environmental system management and the intentions of ISO 14001, this section discusses, in broad terms, the six clauses of ISO 14001. It explains their intended purpose, prior to a fuller examination in Chapter 3.

In the UK, the full title of ISO 14001 is 'BS EN ISO 14001 – *Environmental Management Systems – Specification with Guidance for Use*', where the *BS* denotes publication and adoption as a British Standard and *EN* denotes adoption as a

European Standard. Other countries will have similar national standards prefixes.

ISO 14001 requires organizations to identify the environmental aspects of their activities, products or services and to evaluate the resulting impacts on the environment, so that objectives and targets can be set for controlling significant impacts and for improving environmental performance.

ISO 14001 specifies the environmental management system requirements that an organization must meet in order to achieve certification by a third party – the certification body. (The Standard was specifically designed to be an auditable standard leading to independent certification.)

The requirements of ISO 14001 include:

• Development of an environmental policy

• Identification of environmental aspects and evaluation of associated environmental impact

• Establishment of relevant legal and regulatory requirements

• Development and maintenance of environmental objectives and targets

• Implementation of a documented system, including elements of training, operational controls and dealing with emergencies

• Monitoring and measurement of operational activities

• Environmental internal auditing

• Management reviews of the system to ensure its continuing effectiveness and suitability

The informative Annex A of ISO 14001 contains additional guidance on the use of the Standard. Annex B contains a matrix of the linkages and cross references between ISO 14001 and ISO 9001 and will probably be of interest to those organizations who wish to combine these two separate management systems. This is addressed further in Chapter 5.

It should also be noted that ISO 14004 (*Environmental Management Systems – General Guidelines on Principles, Systems and Supporting Techniques*) was published in 1996 and gives general guidance for implementing organizations. (See Appendix III.)

However, before venturing into the clauses, there is one area of implementation which appears to have been left out of the Standard – and can be considered to be clause 4.0 (if one wishes). This is the Preparatory Environmental Review (PER). A *preparatory environmental review* is an investigative exercise – a structured piece of detective work – which identifies all of the organization's environmental aspects (and is addressed in more detail in Chapter 3). Strangely, this initial step is not mandatory and cannot be assessed during the certification audit (see Chapter 4) and yet, if it is not performed, the whole environmental management system may not be soundly-based. An organization may have a clear vision of where it would like to be in terms of future environmental performance. However, unless this 'snapshot' of current performance is undertaken – the PER – the organization may act in an unfocused manner and not achieve this goal.

So unless an organization knows where it is *now* with regard to its interaction with the environment, it may not be able to move in the correct direction (forward) in controlling and minimizing its environmental impacts. It is only after performing a preparatory environmental review that a meaningful environmental policy, with proper and relevant objectives and targets, can be set out.

The six clauses of ISO 14001 are titled as follows:

4.1	General
4.2	Policy
4.3	Planning
4.4	Implementation and Operation
4.5	Checking and Corrective Actions
4.6	Environmental Management Review

These are now discussed with emphasis on the concepts of the clauses.

Clause 4.1: General

The Annex applicable to this clause of the Standard describes, at length, the intended purpose of the environmental management system (that is, an improvement in the environmental management system is intended to show an improvement in environmental performance). Additionally, because the Standard does not indicate what level of maturity, or implementation, is required of the environmental management system for it to be assessed by an independent third-party certification body, indications are given in this clause.

The words used in the Standard refer to 'establishment' and 'maintenance' Thus, there must be some objective evidence of the system being operated and reviewed.

The review can take the form of monthly progress meetings, corrective actions from audits and of course audits themselves. Audits are evidence of reviewing; that is, asking the question: 'Are the planned activities of the organization occurring in practice?'

It would be very prudent for an implementing organization to offer evidence of – at the very least – one or two audits of a significant impact over a period of two to three months. Keeping documentary evidence of management reviews (even if not the fully formalized 'Management Review' that clause 6 of the Standard refers to) and having some evidence of awareness training for those staff who control those environmental aspects that have a highly significant environmental impact, would also be wise steps to take.

Clause 4.2: Policy

The intention of this requirement of the Standard was that by making the organization's environmental policy available to the public, the organization was very clearly setting highly visible environmental objectives. By this, the organization demonstrated commitment and accountability which could be verified, examined, and even criticized, if it failed to deliver the promises made.

Thus, the policy was intended to be the main 'driver' of the environmental management system and all other elements of the system would follow on naturally from it.

The intended audience for the environmental policy was the lay public, who would see that the organization is taking environmental responsibility. In effect, the organization was stating 'We have nothing to hide'.

In fact, the concept of public availability is open to many interpretations and in reality it is only the stakeholders who are interested in the content of the environmental policy (rather than the general public). Such stakeholders are:

- The consumer of the organization's products or services

- The neighbours bordering the organization

- The shareholders

- The lending institutions

- Insurance companies

- The employees

The standard writers also wanted the policy of the organization not to be in conflict with higher level corporate strategy so that coherence of policy could be seen to flow throughout the wider organization. This was to ensure that no policies of higher-level management would prevent the policies of lower-level management being effectively carried out, for example at the site level.

The environmental policy must also be reviewed by top management. This was to ensure that the ultimate responsibility for, and commitment to, an environmental management system belongs to the highest level of management within the organization.

Because the environmental policy documentation is so fundamental, the meaning of the words on it must be clear and very relevant to the organization's activities. This is explored in depth in Chapter 3.

Clause 4.3: Planning

In any walk of life, unless we plan for an event, it may never happen or alternatively it will happen in an uncontrolled, haphazard fashion and the desired result of the event may not be what we want. Thus planning for an environmental management system must be undertaken in a structured fashion. The purpose of a 'planning' sub-clause is to act as a guide.

Clause 4.3.1: Environmental aspects

The intent behind this sub-clause was to ensure that an organization had the capability and mechanisms to identify continually any environmental aspects it had, and then to attach a level of significance to those aspects in a structured and logical way.

Because the environmental behaviour of a supplier, or indeed a customer, could well turn out to be not of the same level of responsibility exercised by the implementing organization, such 'indirect' or remote activities may well be of far more significance than that of the 'direct' impacts of the organization itself. It therefore makes sense for an organization to include such 'indirect' environmental aspects within its system and, using the same methodology, attach a level of significance.

Clause 4.3.2: Legal and other requirements

This sub-clause requirement was included in the Standard because it was recognized by the authors of ISO 14001 (ISO Technical Committee 207 – TC 207) that an organization could fall down on its environmental performance if it did not possess sufficient knowledge of applicable environmental laws, or codes of practice, within its industry sector. These codes of practice are the 'other requirements'.

First and foremost, an organization must comply with local and national legislation. By definition, legislation exists to control significant environmental impacts, otherwise the legislation would not have come into being. Thus, because of this implied significance, compliance with legislative requirements is the baseline for certification to ISO 14001. Many industrial sectors have membership bodies (such as the CIA – the 'Chemical Industry Association') who issue codes of practice to their members. These are generally guidance notes to ensure 'best practices' are followed. They tend to emphasise health and safety, and increasingly, environmental issues. One can be sure that, if a code of practice exists, there must be sound reasons why it was written. An organization would be expected to comply with it or to demonstrate compelling reasons why such a code had been disregarded.

Clause 4.3.3: Objectives and targets

Although the organization may have an environmental policy (derived from consideration of its preparatory environmental review) and may have identified those aspects of its business which have a significant environmental impact, it needs to translate such findings into clear achievable objectives, measured by specific targets. In practical terms, each significant environmental impact should have an associated objective (and targets) set against it for the control of, and minimization of, that impact. This then is the intention and purpose of this particular sub-clause.

Further, although the overall aim of the environmental management system should be continual improvement, not all the objectives have to relate to immediate improvements in environmental performance. However, such objectives must ultimately demonstrate overall improvement in performance. For example, there could be cases where improvements are identified as an objective but this objective may only be realized several years in the future when, perhaps, either investment or technology currently being developed will be available.

So there may well be a range of objectives: some completed in the short term, others being reached only in the longer term.

Clause 4.3.4: Environmental management programme(s)

The purpose of this sub-clause is to ensure that the organization has allocated responsibilities and resources and set time-scales for ensuring that the activities described in the preceding sub-clauses will happen as planned, and also that new activities will be subject to such environmental management controls. This ensures that the environmental consequences of any new developments are considered at the earliest possible stage. To allow such a programme to be monitored, it therefore makes sense for the organization to document and make visible, and available, such a plan or programme to all involved employees.

Clause 4.4: Implementation and operation

This, the longest section in the Standard, has no less than seven sub-clauses, and is written to enable an organization to operate an environmental management system to the requirements of the Standard on what might be referred to as an everyday basis.

These sub-clauses cover:

• The requirements of personnel responsibility and training

- How communications should be handled both internally and externally to the organization

- What documentation should be in place for personnel to refer to

- What is required to control processes with significant environmental impacts

- How to act in the event of an environmental incident or emergency

The sub-clauses are described below.

Clause 4.4.1: Structure and responsibility

This sub-clause was included to ensure that personnel are assigned specific responsibilities for a part, or parts, of the environmental management system and have a very clear-cut reporting structure (with no ambiguities). For example, when monitoring emissions to atmosphere, it should be clear who actually performs this task, with contingency plans for responsibility if the named person is away ill or on vacation.

History shows us that when an event turns into a crisis, the cause is often due to the fact that no one individual takes ownership of a problem. Some crises are caused by mismanagement in the form of disregard for safety procedures, for example, but in the majority of incidents, the people concerned did not know they were responsible. Either there was no clarity of roles or interfaces of responsibility were blurred.

Therefore, with this in mind, the sub-clause also requires that top management appoint an individual to be the 'management representative', with specific ownership for the well-being of the environmental management system and co-ordination of all environmental activities.

Clause 4.4.2: Training, awareness and competence

This sub-clause is designed to enable an organization not only to identify training needs, as appropriate, but also to measure the success of that training. All individuals need some form of training to enable them to perform a new task. However, some training does not include equipping the trainees with awareness of the consequences if they fail to perform that task adequately.

It should be understood that there is a big difference between training and awareness. *Awareness* is the product (or end result) of any training given – if personnel are more aware of the consequences of their actions they are far more likely to follow procedures.

Competence is usually a term describing an individual's capacity to absorb training and to apply the resultant awareness to the tasks that they perform. So, although several individuals may appear to be equally receptive during a training session, there may well be big differences in performance when the knowledge from the training is put into practice. Some will do better than others; that is, they will be more competent – and this is an area on which this sub-clause requires the implementing organization to focus.

Clause 4.4.3: Communication
It was believed by the authors of the Standard that means of communication, both internally and externally, are extremely important and that, if not formally addressed, may have negative effects on the success of the environmental management system.

Internal communications are necessary for information flow, particularly to departments that may not be at the forefront of the environmental management system. For example, the accounts department is an area which might control environmental issues through payment regimes. Cross-departmental barriers can exist in any medium-sized organization. The concept of 'the internal customer' is still in its infancy and internal information flow can still be poor. External messages are just as important (such as getting the message across that the organization is taking steps to be environmentally responsible). Such communication needs to be co-ordinated to get across a consistent message to the media, suppliers and customers, as well as to other stakeholders.

Clause 4.4.4: Environmental management systems documentation
For a system to be audited, there must be a minimum level of documentation available to demonstrate that the system exists and can be followed through by anyone who wishes to do so. This 'hard' evidence can be in the form of electronically held data, of course.

The sub-clause itself uses phrases such as 'describing the core elements' and 'provides direction', which indicates that a top-heavy documented system is not the aim of this sub-clause. There is encouragement from the accreditation bodies and certification bodies alike to ensure that an environmental management system is not too focused on documentation alone. A balance must be struck between failure to documents essentials and a bureaucratic system that does not add any value or meaning to the system. For example, where ISO 14001 calls for a 'written' procedure, then clearly there must be such a document to achieve compliance with the Standard. In many other parts of the Standard, documenting a reference to existing documents – training manuals, Quality Assurance Procedures, machinery/plant operating instructions etc. – will suffice.

Additionally, smaller organizations should note that if the environmental impacts of the organization are complex, then this complexity will dictate the level of documentation required to control and minimize such impacts. Straightforward environmental impacts may only require a modest level of documentation – perhaps backed up by evidence of high levels of awareness and competence being demonstrated by operating personnel.

'Providing direction' can mean that the documentation merely signposts where records, procedures etc. are kept. For example, if an organization has an established quality assurance system, then it will already have appropriate procedures for identifying training needs and where evidence of training (in the form of records) is filed. Such a procedure may then be referenced in the environmental management system. There is no need to duplicate what may be a perfectly adequate procedure. Again, accreditation bodies and certification bodies encourage such system integration. Showing cross-references to other management systems is also encouraged (for example, to occupational health and safety management systems).

It would be prudent for a simple matrix to be drawn up listing all the clauses and sub-clause headings of ISO 14001, and against each item on the list, the corresponding procedure title of the organization. This would show in a clear and visual manner whether all the requirements of the Standard had been addressed.

Undoubtedly this makes life easier for the external auditor when ensuring that the client has addressed all parts of the Standard, but the primary purpose of such a matrix is to help the client be confident that all parts of the Standard have indeed been addressed. The requirement that all clauses are addressed is mandatory for successful certification.

Clause 4.4.5: Document control

The purpose of document control in any management system is to ensure that when, for example, an operator follows a procedure, that procedure is the most up-to-date one available, and that an out-of-date procedure cannot be followed accidentally. In an environmental management system, following an outdated procedure could lead to adverse environmental consequences, so some method must be in place to control documentation.

In the Annex to the Standard, the relevant clause encourages organizations not to have a complex document control system. This is also the stance taken by accreditation and certification bodies.

Clause 4.4.6: Operational control

The purpose of operational control is to ensure that those environmental aspects that are deemed to be significant (as identified earlier in clause 4.3, Planning) are controlled in such a way that the objectives and targets have a fair chance of being achieved.

Such controls will invariably be described and documented in procedures but need only be appropriate to the nature, complexity and degree of significance of the function, activity or process that they address. Such controls should include both direct and indirect environmental impacts.

Clause 4.4.7: Emergency preparedness and response

The intent behind this sub-clause is that an organization must have in place plans of how to react in an emergency situation. Waiting until an emergency occurs and then formulating a plan is plainly not a good idea. The emergency plans or procedures may not work in practice, and this failure may lead to an environmental incident.

Thus it makes sense to identify the potential for an emergency, identify the risks and put plans in place to prevent and mitigate the environmental impacts associated with such an emergency. Several options are open to organizations, ranging from, at the simplest level, listing competent personnel who can be contacted (with alternatives) in the event of an out-of-hours emergency situation, to predicting worst-case scenarios that might involve serious pollution and perhaps loss of life. Such plans should be periodically tested, in the absence of genuine emergencies, to verify that they will work in practice.

Clause 4.5: Checking and corrective action

In any organized venture operated by people, experience again tells us that, without proper supervision or management, standards will fall. People will make mistakes under pressure and some individual will find an easier way of performing a task and keep it hidden as a token of individualism. Somewhere along the way, the original common goal may be forgotten. Unhappily, that is the way of the world and no matter how robust our management systems are, and how motivated our personnel are, it would be very foolish not to install some form of monitoring and checking and, if circumstances dictate, to take corrective actions. Corrective actions should, if undertaken, prevent a similar event happening again.

This, then, is the intent of this clause: to ensure that the organization has a system that is robust enough so that even if an event does not take place as it should, this 'nonconformance' is recognized within the environmental management system and appropriate corrective actions are taken.

Clause 4.5.1: Monitoring and measurement

A programme or activity for environmental improvement cannot be said to be achieving anything unless the starting point is known, the objective and target are defined, and progress in between start and finish is somehow measured. Thus, this sub-clause requires an organization to monitor and measure its environmental targets at regular intervals. Unless there is such regular monitoring, an environmental objective may not be achieved. Furthermore, the organization may not recognize this as a problem, nor take the necessary corrective actions.

Of course any equipment used during the monitoring process must be reliable so that personnel using such equipment are confident that the readings shown are accurate. Such confidence in measuring equipment can, of course, be obtained by a systematic programme of calibration, and this also is the purpose of this sub-clause.

Clause 4.5.2: Nonconformance and corrective and preventive action

This is an extension of the opening paragraph, which described corrective action in general terms. Nonconformances in the system must be recognized and acted upon. The root cause should be investigated and controls put in place to make sure the nonconformances do not happen again. Although this is the overriding purpose of this sub-clause, care must be taken to ensure that the corrective actions that are taken by the organization are commensurate with the environmental impact encountered and that committing excess time and resources to problems of a low magnitude is avoided.

Clause 4.5.3: Records

The purpose of this clause is to ensure that the organization keeps records of its activities. For example, in the event of a dispute with a regulatory body, not having records to demonstrate compliance with discharge consents (in the form of independent monitoring and measurement data) could spell trouble for the organization. A potentially heavy fine may be reduced if objective evidence in the form of records is produced which demonstrates – if not absolute control – then due diligence. It therefore makes sense for the organization to decide which records it needs to keep, and for how long, commensurate with the risks involved if they did not keep such records.

In any event, legislative requirements will dictate that some records are kept for minimum specified time periods.

Clause 4.5.4: Environmental management system audits
Internal audits are now an established management tool in many businesses. The concept of *self-policing* is recognized as an improvement mechanism by organizations with any form of management system. Environmental management systems are no different and this sub-clause requires that such audits are carried out.

Accreditation bodies insist that third-party certification bodies must determine the amount of reliance that can be placed upon the organization's internal audit. Such audits should be carried out in much greater depth than the external assessment body could hope to achieve and, indeed, this is an area upon which the certification body places much emphasis. The completeness and effectiveness of internal audits are major factors in demonstrating to the certification body that the environmental management system is being well managed.

Clause 4.6: Management review

The purpose of this clause is to consider, in a structured and measured way, all of the preceding steps that have been taken by the organization, and to ask fundamental questions such as:

- Is the organization doing and achieving what has been stated in the environmental policy?

- Are objectives and targets that are set for environmental performance being achieved?

- If objectives and targets are not achieved, why not?

- Are appropriate corrective actions taking place?

These questions, and more, should be asked by top management. The ideal vehicle for such an inward-looking review is a formalized management review with an itemized agenda, minutes being taken, and a report being issued to all interested parties.

A guideline for the time interval between reviews is 3 to 6 months in the early stages of implementation followed by annual reviews once the system becomes more mature. In reality, the time intervals should be determined by events.

For example, if it is found that the objectives and targets are being met, with very few exceptions, then the organization is well on its way to minimizing its

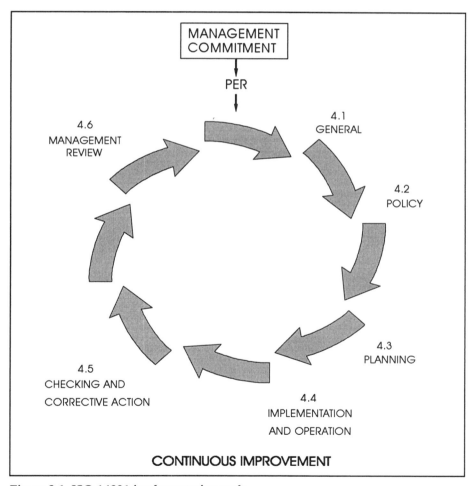

Figure 2.1 ISO 14001 implementation cycle

environmental impacts and thus complying with the intended spirit of the Standard. Management reviews need only then take place annually. Conversely, many failures to achieve objectives demonstrate that the environmental management system has weaknesses and that more frequent management reviews should be held. Figure 2.1 illustrates the cyclic nature of continuous improvement required by the Standard, each 'cycle' culminating in the occurrence of a management review.

Summary

This chapter has looked at the concepts of environmental control. The point was made that being emotive about reducing environmental impacts may cause us to use a 'scatter-gun' approach to environmental management rather than one more focused on meaningful improvements in environmental performance by using the concept of significance.

This concept of 'significance' lies behind the wording of the Standard – its intention or 'spirit'. The actual words and phrases used are the product of the many international parties involved during the process of writing the Standard, who ensured that clarity was paramount. This clarity meant that the Standard could be audited against with some measure of consistency at an international level. (The Standard also had to be written to be generic so as to apply to any organization.)

The clauses of the Standard were introduced in this chapter and so was the importance of a preliminary review – the PER. Chapter 3 now extends the reader's conceptual knowledge by studying the clauses and sub-clauses of the Standard in detail.

Chapter 3

Implementation of ISO 14001

Introduction

Chapter 2 introduced the fundamental concepts of ISO 14001 – why certain elements and requirements have been put into the Standard and why it is structured in the way that it is. This chapter builds up a more complete picture for the reader by addressing the detailed requirements of the Standard, clause by clause.

As before, illustrative examples are provided where appropriate. The intention of these examples is to stimulate thought as to the best way to interpret the particular clause of a generic standard for the reader's own organization.

First, a word is needed here on the concept of *commitment*. It is a word that is often misused. The effective implementation of the system can only be achieved if there is absolute commitment from top management. Such commitment includes the allocation of management time for the implementation phase as well as funding. Costs will be incurred during implementation and these must be budgeted for. Without a high input of such resources at the start of the implementation phase, the system will flounder and collapse.

This show of commitment is really no different to any other undertaking by the organization. If the junior personnel and staff see no commitment from the Managing Director to a new product or process, or development of a new market, they are hardly likely to be enthusiastic themselves.

The resources of the organization will be stretched for some time, even if external consultants are used. Extra training will be required. Individual staff may have to perform tasks above and beyond their primary tasks. Therefore, management must ensure that this happens with some show of interest and enthusiasm and that the implementation process comes, temporarily at least, at number two or three on the business priority agenda: number one priority being maintaining a profitable business, of course.

How does an organization start?

Commitment is required from top management before the organization begins to implement an evironmental management system. The Standard actually refers to structure and responsibility, and is really requiring a high level of management commitment (sub-clause 4.4.1 *Structure and Responsibility*, described later in this chapter).

Organizations are generally unsure of how long they can give this commitment for. In truth, it is very difficult to give an optimum figure for the required length of time an organization will need for the implementation phase. However, based on the experience of both large and small companies, twelve months is too short a time period, whereas three years is too long. Eighteen months seems to be the 'norm'.

The reasons for this particular figure of 18 months are very much to do with people. Much of the success of any project's implementation is due to personnel having enthusiasm for the project's successful conclusion. People make it happen. However, time is required to change habits, to absorb the new culture and ways of working, to turn a new concept into just part of everyday working practices.

In the early stages of a new environmental management system there is nothing very tangible for personnel to see. Therefore, if the time-scale is too protracted so that personnel cannot see an end or a goal being reached, then invariably enthusiasm wanes and the project begins to go 'off the boil'. Cynicism may set in, personnel cannot see certification ever being gained, and the project fades away. Enthusiasm will never again be rekindled. So it appears that 18 months (and at the outside, two years) is the time-scale to aim for. The organization should 'switch on' the resources, if necessary, to reach this target.

It may be that an organization has one or more existing management systems. If so, it is possible – and perfectly acceptable from certification bodies' point of view – to use elements of such systems that already exist within the organization as part of the new environmental system (see also Chapter 5). Such existing systems could be quality assurance, occupational health and safety, and so on.

The preparatory environmental review will be discussed in detail now, prior to going into the details of the six main clauses of the Standard:

4.1	General	
4.2	Policy	
4.3	Planning	
4.4	Implementation and Operation	
4.5	Checking and Corrective Actions	
4.6	Environmental Management Review	

A description of these clauses is given in Chapter 2.

The preparatory environmental review (PER)

An organization considering the implementation of ISO 14001 should decide whether or not it needs to perform a *Preparatory Environmental Review (PER)*.

This review is not mandated by the Standard, but in the Annex to the Standard (section A.3.1) it is suggested that an organization with no pre-existing environmental management system should establish its initial position, with regard to the environment, by a review.

As stated in Chapter 2, the preparatory environmental review is not an auditable item on the external auditor's check list but, by examining the review, the auditor will get a measure of the environmental competence of the organization and an indication of the level of understanding of environmental issues by the organization. In short, a 'feelgood' factor will be gained which can only assist in the smooth conduct of the certification audit (see Chapter 4).

Furthermore, if an organization has been operating an informal environmental management system for some time, it will in all probability have a sound knowledge of its environmental aspects, acquired by the experience of operating such a system. This will be true even if such an unfocused system does not meet the requirements of ISO 14001 itself. In essence, it has performed the equivalent of a preparatory environmental review.

Conversely, an organization new to the principles of environmental control or an environmental management system is well advised to take the opportunity to perform a structured review and to base their environmental management system upon the results of that review. If the decision is taken to move straight into implementation, there is a large body of evidence (from the certification bodies) that has shown that such systems are seriously flawed. The organization has had to rethink its strategy – with resultant setbacks in the certification time-scale.

If an organization decides to undertake this preparatory review, there are two options open:

1   Perform the preparatory environmental review using internally available resources.

2   Perform the preparatory environmental review using external consultants.

These options are considered below.

Internally available resources
Performing the preparatory environmental review using internally available resources has its merits in that the organization can use personnel experienced in the operations of its processes and, of some importance, costs can be somewhat better controlled.

Several options exist for this approach. One option is to send out questionnaires to each department head, requiring those individuals in charge to complete a series of questions, including:

- What materials are used

- What quantities of materials are used

- How much energy is used

- The amount and type of waste streams

- Possible emergency situations

- Abnormal situations (frequency of start ups and shut downs, planned maintenance and breakdowns)

- Any areas of training required

This will form a meaningful exercise by establishing the baseline to work from after analysis by management.

Possible flaws in this approach are that staff employed to perform this task (perhaps as an extension of their existing duties) may not have the necessary expertise to carry out a meaningful review. (Careful review of the answers in these completed questionnaires will indicate the level of training required in the personnel completing them!) Management tools such as 'brainstorming', although they are of value, will not give the same answers as hard data collection and some investigative detective work.

On ocassions, organizations have made the mistake of basing their preparatory review on environmental projects that are currently up and running. The rationale is that if such projects are current, then they must be important, must focus on the significant environmental impacts of the organization and, there-fore, must be a sound basis to start from. Unfortunately, although such projects may have been started with the best of intentions, they may be based upon previous initiatives (for example, a project that was topical at the time, or a project that looked easy to fulfil). The project may have been used to give credence to an individual or the organization during a marketing initiative at that time. Or again, it may have been a project which tied in with everyone's work schedules and was easy to manage, and with which everyone was comfortable because of the feeling that 'they were doing something for the environment'.

Such projects may well have been reducing environmental impacts and this is no reason to abandon them. Unfortunately, because of the haphazard nature and methodology of such projects they will, in all likelihood, not be focusing on significant impacts – fundamental to ISO 14001 philosophy.

It is essential if following this option that at least one senior manager in the organization has environmental expertise. If the expertise is not available within the organization, suitable existing staff might be trained by external consultants. This leads on to the second option.

External consultants
Quite often the individual chosen to lead the environmental management team and implement ISO 14001 is the quality assurance manager. The reasons for this choice are sometimes baffling and no disrespect is intended here for those quality assurance managers who have taken on this task and are very competent. There are of course commonalities between quality systems and environmental systems. A quality assurance manager will be well versed in operating to documented management systems: the concepts and day-to-day administration; the requirement for self-policing (auditing); the value of reviews; and a corrective and preventive action system to allow improvements to occur. But this does not necessarily equip the manager in question with the knowledge and skills in environmental issues required for ISO 14001. That said, the requirement for in-depth environmental knowledge from within the organization need not be onerous, and much of course depends upon the complexity of the organization's environmental aspects. A quality manager from a scientific background will probably be able to grasp environmental concepts more easily but, nevertheless, the organization must ask itself whether this is the correct choice of individual.

Knowledge of environmental legislation is an area where there could be a shortfall of information available from within an organization. In many countries, including the UK, there are organizations whose business is keeping up to date with legislation and providing such information for a fee. This has immense value because it enables an organization to keep up to date with legislative issues.

However, to commence from a standing start into the complex world of environmental issues and perform a meaningful preparatory review is something not to be undertaken lightly. In such cases, the organization is well advised to use the services of an environmental consultancy and ask them to perform a preparatory review prior to ISO 14001 implementation.

Whichever route is taken, a well-executed preparatory environmental review will generate a 'specification' for the organization in the form of a report setting out what it needs to do. This will form the foundation for deriving a meaningful environmental policy and developing a robust environmental management system, capable of demonstrating environmental performance improvement.

A typical format of a preparatory review is given below – in the form of a list of questions – and is an example of the approach required. It is followed by an example of a typical summary report that an organization might receive from a review performed by an external environmental consultancy.

This fictitious preparatory review and report is illustrative only and is not intended to be exhaustive.

Steps to take – the checklist approach for a PER

The approach taken in the example of a PER below follows the informative advice supplied with the Standard in Annex A.3.1 and considers four key areas:

1   Legislative and regulatory compliance

2   Evaluation of significant environmental aspects

3   Examination of existing environmental practices and procedures

4   Assessment of previous incidents

Each of these areas is considered in turn.

Legislative and regulatory compliance

A fundamental requirement of ISO 14001 is that the organization complies with environmental law as a minimum standard. The review should identify which areas of the organization are covered by which laws. Any areas where there are breaches of legislation should be set as priority action areas. Typical questions to ask are:

a)   *Is all existing legislation being adhered to?*

Air emissions:
If there is an authorized process, are the requirements for measuring, monitoring and recording being complied with?

Solid waste to land:
Is a waste management licence required for the site? If solid waste is taken away to landfill, do the operators of the landfill site have a licence? Does it cover the particular waste that is being removed? Does the carrier of the waste need a licence?

Water:
Is there any groundwater extracted on site? Is a licence required for this? Is any effluent discharged to streams, rivers, local authority sewage systems? Is a licence to discharge required? If so, are the conditions being met?

Other:
Does the site have any obligations to comply with any town and country planning consents or building regulations?

b)   *Is there any forthcoming legislation which may affect the business?*

For example, there may be legislation in the draft stage that, if enacted, could put the organization under heavy financial strain to comply. Therefore some investigative work is required.

c)   *Have there been previous incidents of breaches of legislation?*

Has the organization been prosecuted for any breach of environmental legislation? This may point to areas of weakness in the management system.

Evaluation of significant environmental aspects

An examination of an organization's environmental aspects is a key requirement of an environmental management system because unless all are identified, a potentially significant impact may be overlooked.

The following range of questions will establish a baseline from which the evaluation of the environmental impact can be determined (see clause 4.3 *Planning* later in this chapter).

Aspect	Questions
Raw materials	What materials are used and in what quantity?
Water	What quantities are used? Is any recycled?
Gas/Oil/Electricity	What costs are incurred?
Solid waste	What quantities are sent to landfill?
Effluent	What amounts are sent to sewer?
Emissions to atmosphere	What amounts are emitted?
Hazards	Are all hazards known about chemicals on site?
Transport and distribution	Does the organization know the amounts of fuel used during distribution of its products?
Past history of the site	Is this known? (Previous use of the site may have contaminated the land, making it unsuitable for certain future uses.)
Locality	Who are the neighbours? Is it an urban/industrial/rural area? Are there any special sites nearby for conservation of natural habitats? Are there any high amenity areas nearby?

Examination of existing environmental practices and procedures

There are many aspects to environmental management within an organization, especially management commitment. Such commitment is demonstrated in the environmental policy and documented procedures which are necessary to ensure such policies are known, understood and followed throughout the organization. The following checklist illustrates the scope of the management system that needs to be in place, with some typical records that need to be retained to demonstrate a minimum level of an environmental management system.

Is there a documented:

* Environmental policy

* Management structure

* Management system

Are the following environmental records kept:

* Discharge consents

* Process authorizations and amendments

* Duty of care waste transfer notes

* Planning consents

* Emission monitoring

* Discharge monitoring

* Maintenance and calibration of pollution monitoring and control equipment

Further, what is the level of awareness of environmental issues possessed by senior management, middle management and supervisors, technical and commercial staff, and production operators?

Assessment of previous incidents

The preparatory environmental review should also include an assessment of previous incidents – under both abnormal and emergency situations.

* Abnormal conditions include start-up and shut-down of continuous processes.

* Emergency conditions include fires, floods and chemical spillage.

Therefore, typical questions to ask are:

- Has a formalized risk assessment been carried out?

- Are there emergency plans in place – especially concerning a major spillage or fire?

- Are staff trained in operating to such plans?

The outcome of the review should be a report which, as mentioned earlier, should form the basis of a specification. This specification, if adhered to, will provide a good foundation for the environmental management system.

The preparatory environmental review report

This report typically describes the scope of the audit and may commence with such a statement as:

> *'The review consisted of a physical on-site inspection of all external boundaries, drainage systems and processes carried out. Operating staff were interviewed and internal records consulted. Plans and maps from both internal and external sources were consulted and the regulatory framework was examined through discussions with the appropriate national and local enforcement authorities.'*

The report could then go on to indicate those items that have the highest environmental impact and should be addressed as a matter of urgency.

Key:	
P = priority **C** = possibility of complaint **G** = good practice	

Emissions:

P	Comply with the requirements of the coating process authorization.
P	Maintain documented monitoring requirements of visual and olfactory survey as required by the authorization to operate.
P	Compile inventory of legislative and regulatory requirements.
	Continue to research and review potential use of alternative coating processes to reduce VOC emissions (e.g. use of water-based paints and powder coatings).
C	Develop codes of practice for sub-contractors on site during the present building programme to reduce fugitive dust emissions – and thus reduce complaints.
G	Monitor and control external noise emissions and intrusive lighting, as appropriate, to reduce disturbance to adjacent sensitive amenity areas.

Discharges:

P	Ensure continual compliance with trade effluent consents to discharge.
C	Determine the precise route and discharge outfall of the surface water should in order to establish the potential environmental impacts of inadvertent releases of water contaminated by chemical paint spillage.
G	Make sure that all the site drainage points are identified and that their routing to the receiving waters is known, especially to the owners of the route between the site boundary and the receiving water.
G	Carry out regular inspection and maintenance of surface water interceptors; this should be the subject of an operational control procedure.

Waste Management:

P   Monitor and ensure regulatory compliance with regard to maximum allowable storage requirements of dry wastes.

P   Ensure all waste disposal facilities are appropriately licensed.

G   Clearly label and designate waste collection areas to provide unequivocal waste segregation centres in order to prevent mixing of wastes (for example, metals) and to encourage on-site waste recovery and recycling.

Storage Facilities:

P   Establish specific bunding requirements for drummed materials at 110% capacity.

G   Provide adequate labelling of hazardous raw material stores and segregation from waste materials.

G   Monitor storage of waste closely. Consider potential for cross contamination, corrosion of containers and possible deliberate spillage by vandals. Use waterproof and wind-resistant labelling for containers.

G   Perform regular inspections and maintenance of bulk process, in-line and chemical storage tanks.

G   Develop and implement an energy conservation programme – heating and lighting.

G   Limit the wide variety of packaging used to encourage future segregation and recycling.

Suppliers:

P   Increase environmental probity of suppliers through site inspections, questionnaires and awareness seminars.

G   Incorporate environmental components in materials specification and supplier sourcing.

Customers:

G   Provide customers with environmental performance criteria of finished products; increase their awareness of their duties & obligations for safe & appropriate disposal of products.

Having performed the preparatory review, and acted upon the recommendations, especially treating those marked 'P' as being of a significant nature, the organization should now be in a position to prioritize an action plan and begin implementing the Standard.

Clause 4.1: General

This clause of the Standard states that *the organization shall establish and maintain an environmental management system, the requirements of which are described in the whole of clause 4.* (Clause 4 actually encompasses all the requirements of the Standard!)

As discussed in the previous chapter, this clause of the Standard is deceptively brief. However, referring to the related part of the Annex within the Standard (A.1 *General Requirements*) several fairly wordy, descriptive paragraphs are used to explain that the intention behind implementation of the Standard is to give improvements in environmental performance. By the continuous process of reviewing and evaluating, an environmental management system will be improved with the intended result of improving its environmental performance.

There is no real guidance as to how mature an organization's environmental management system needs to be to qualify for certification. The phrase 'establish and maintain' does not make this clear. For example, how long must the system have been established, and how much documentation needs to be generated to demonstrate a history of implementation, before the third party assessment? Further guidance must be sought in the relevant section of the *Accreditation Criteria* (see Annex IV) and this is addressed further in Chapter 4.

Clause 4.2: Policy

The Standard requires that *top management shall define the organization's environmental policy and ensure that it:*

- *is appropriate to the nature, scale and environmental impacts of its activities, products or services;*

- *includes a commitment to continual improvement and prevention of pollution;*

- *includes a commitment to comply with relevant environmental legislation and regulations, and with other requirements to which the organization subscribes;*

- *provides the framework for setting and reviewing environmental objectives and targets;*

- *is documented, implemented and maintained and communicated to all employees;*

- *is available to the public.*

The policy should, therefore, be relevant to the significant impacts of the organization and should focus on them. For instance, a plastics processing company should not focus its policy on saving raw materials by recycling polystyrene drinking cups in the canteen yet ignore highly-significant waste from the manufacturing processes.

There are conflicts that need to be considered when producing a policy statement for any organization and a balance must be achieved. On the one hand, the policy must be specific, yet general enough for the public (remember that the Standard requires public availability) so that a 'lay' person could read it and identify the processes/products of the organization and what the organization is planning to achieve with respect to its environmental issues. On the other hand, the policy must not be so specific that it becomes outdated too quickly, or that the objectives and targets are too exacting and hold the organization to promises it cannot keep.

One solution adopted by some organizations is to produce a meaningful policy but without details of objectives and targets. A more detailed document is available separately upon request, via a different document which is perhaps reviewed every three to six months whilst the policy withstands scrutiny from year to year. The organization should have nothing to hide and everything to gain by demonstrating environmental responsibility.

Commercial confidence must also be considered during policy writing and it was never the intention of the Standard to compromise an organization by forcing it to reveal sensitive information that could be used by competitors. As a guideline, the certification body is looking for a balanced statement that can be audited against. Therefore, if an organization states that every member of staff and all contractors have been trained in environmental awareness, the auditor will look for evidence of this.

A good guideline is to strike a balance in the policy. Include longer term and short term objectives as well as highly specific and broader objectives.

Where an organization is part of another organization or corporate body, it should ensure that its site environmental policy does not conflict with any statement in higher-level policies of the wider organization – and yet be very relevant to the site. The Standard focuses on this requirement to ensure that no higher-level policies would prevent the lower-level policy being carried out effectively at the site level.

If, for example, targets are set for a reduction of a particular emission to atmosphere, the auditor will need to see evidence of such reductions. The decision-making processes used by management in setting particular targets will also be examined.

Figure 3.1 illustrates a rather weak environmental policy that raises more questions (as in the notes at the side) than it answers. Could it be understood by the stakeholders? Could it be audited against by both internal and external auditors?

In addition, this policy does not address all the requirements of the clause of the Standard! It does not mention compliance with legislation, commitment to pollution minimization and does not appear to relate to a preparatory environmental review. The organization manufactures dustbins yet the policy does not reflect the particular environmental aspects and impacts of dustbins. Finally, who is M. T. Cann?

A more balanced policy statement by *Premier Dustbins* is illustrated in Figure 3.2.

Environmental Policy Statement

Premier Dustbins

Premier Dustbins will adopt an integrated approach to all environmental issues in order to achieve best international practices.

This will involve the management of:

* *The impact of our processes on the environment*
* *Control of substances hazardous to health*
* *Relations with our suppliers*
* *Product and process development*
* *Product liability*

Commitment to this integrated approach must be an integral part of business unit strategy.

Each business unit will assess its current position and then formulate plans and budgets to achieve the standards required.

Improvement plans will be regularly reviewed and audited

Signed

M .J. Cann

Questions the auditor would ask:

What does this mean? What are these practices?

This is commendable - but is more occupational health and safety related.

This is unclear in meaning.

What is required of personnel to achieve this?

What are these standards?

Will any corrective and preventive measures take place?

Figure 3.1 Environmental policy statement

Environmental Policy Statement

Premier Dustbins

Premier Dustbins recognizes the need to manage the impact of its processes and business on the environment.

To manage this impact, the following objectives have been set, based upon examination of the findings of our preliminary review, with detailed targets available to all our stakeholders:

* *To meet best international practices by seeking alternatives to the Zinc Galvanising Process;*

* *To examine the life cycle of our products and seek new manufacturing processes, that allow more use of recycled materials - in particular plastics, cardboard and wood fibre;*

* *To reduce levels of organic solvents used;*

* *To assist our suppliers in formulating their own environmental policies;*

* *To meet all relevant legislative requirements;*

* *To review regularly all environmental aspects of the organization.*

Commitment to the above policy will be demonstrated by a review of improvement plans by top management.

Corrective actions will be taken as required.

M. J. Cann
Managing Director

Auditor's notes:

Very specific to the organization!

Related to the preparatory environmental review

This is very specific to dustbins

Broad objectives

Pollution minimization

A consultative approach to suppliers

Compliance with legislation

Reviews

Corrective actions

Commitment from top management!

Figure 3.2 Revised environmental policy statement

Structure of the environmental policy

The following statements are extracted from other environmental policies and are examples of organizations not really considering the message they are conveying to the stakeholder:

'We will monitor and minimize the impacts of current operations on the environment.'

Does this mean all impacts will be minimized – not the significant ones initially? This could be a tremendous task and may not be workable.

'We will ensure that sub-contractors and suppliers apply equivalent environmental standards.'

Does this mean that all sub-contractors and suppliers have to obtain ISO 14001? If so, by when? How can this be enforced? How can this be controlled? What power does the implementing organization have over a sole supplier who can easily sell their raw materials elsewhere, particularly when losing the organization as a customer would cause the supplier no financial hardship? Again, this may not be a workable objective.

More balanced statements are:

'Through measurement and examination of the impact of its own activities the company seeks to eliminate or reduce the production of pollution.'

'We will work with suppliers and Customers to influence them positively in terms of our environmental policy.'

The policy should not end up as a list of isolated statements which do not have substance. The policy must be very individual to the organization. One test of this is to audit it against the activities of the organization; this is further examined in Chapter 4 and Chapter 7. The policy must always be the driver of the environmental management system, not the passenger!

The following examples are extracted from actual environmental policies. Such examples show a good balance of intent, breadth and scope whilst also being very specific to the organization's business.

A *lighting company* manufacturing light bulbs and fluorescent tubes:

• To consider the environmental aspects of all processes and/or materials, existing and new, and to apply best available techniques not entailing excessive cost (BATNEEC), where appropriate, to help demonstrate our commitment to the prevention of pollution.

• To fund research into longer life products, thus saving valuable and non-renewable raw materials.

• The factory's objectives to reduce the environmental aspects on site will be made available to the public and stakeholders, through the improvement plan and associated projects. These will be implemented through individual improvement meetings.

• To consider the environmental policies of our suppliers when purchasing materials and to encourage the use of recycled and/or recyclable products where practicable.

• To seek out alternatives to the use of rare metals in the manufacturing process.

• To comply with all relevant environmental laws, regulations and corporate policies. The plant will work to achieve whichever sets the highest standard.

A *plastics processor* manufacturing packaging film:

• Within society, we will help to educate all consumers about the benefits and sound uses of plastics films and will be active in external initiatives to develop the efficient use of post-consumer waste.

• We will ensure that all internal scrap is reprocessed and made into saleable product.

• We will reduce virgin material consumption by using recycled or reprocessed material but without compromising customer and consumer quality requirements.

A *brewery*:

* We will minimize use of raw materials which have been grown using chemical agents (fertilizers and insecticides).

* We will only advertise responsibly. Campaigns will not be directed at vulnerable members of society.

* We will examine and reduce distribution costs by increasingly using haulage companies who are considering environmental controls, such as software route optimization and use of 'green' fuels.

A *frozen food manufacturer*:

We will focus on the following environmental issues:

* Reduction of energy costs – by funding research into improved insulation technology.

* Reduction of distribution costs – by consideration of transport routes, types of vehicles, fuel type, use of speed governors and driver training.

* Use of local suppliers of produce, where possible, to reduce transport costs.

* Reduction of food wastage by better shelf-life management.

* Use of brownfield sites rather than greenfield sites for all new freezer depots.

* Formulation of a purchasing policy on seafood derived only from sustainable fish farming practices.

A *service industry* (a local authority):

* Where possible, we will encourage new businesses in the area to use brownfield rather than greenfield sites.

* We will phase out the use of organic weedkillers used in the borough's green amenity areas such as parks and football fields.

* We will rationalize our fleet of commercial vehicles, using the most efficient vehicle for the job and, where possible, use electric vehicles.

A *forestry organization*:

*   Our environmental policy includes a commitment to a program of sustainable forestry management:

    ¤   To preserve the natural bio-diversity of life within the forests.

    ¤   To use planning in all felling and transportation systems to mini-mize disturbance to the environment.

    ¤   To increase hardwoods by modifying forestry practices.

Finally, any organization should ask itself what it is attempting to achieve by the wording of its environmental policy. Minimum compliance with the Standard only? Or can the wording be such as to go one step further? Remember, the policy needs to be publicly available. However, this should be viewed as a positive requirement of the Standard and the opportunity should be grasped to provide more information to interested parties. The policy demonstrates that the organization does take its environmental responsibilities seriously and is taking specific constructive action to manage its environmental impacts. This can only improve the image of the organization in the eyes of the public and other interested parties.

Clause 4.3: Planning

The Standard requires that *the organization shall establish and maintain procedures to identify the environmental aspects of its activities, products or services that it can control and over which it can be expected to have an influence, in order to determine those which have or can have significant impacts on the environment.*

For many organizations, the key difficulty at this stage is the subjectivity associated with the identification and evaluation of environmental aspects. Since this exercise is the fundamental element in establishing a relevant environmental management system, then the methodology used, and the judgement exercised in assessing the significance of impacts is of crucial importance. Of equal importance is the discernment shown in being aware that an environmental aspect can be considered to be insignificant.

The screening mechanism – that is, the criteria for significance used to separate out the significant from the not so significant – must be transparent. This is because it may contain an element of subjectivity – corporate and personal opinions, perspectives and prejudices.

One thing is certain. There must be a range, or differing levels, of significance because clearly, if all environmental impacts are of the same significance, then none are significant! There has to be one impact (or perhaps two or three at the most) that has the highest 'significance', or shares the highest significance. That is the reality and not every environmental aspect is of the same significance at the same time.

Clause 4.3.1: Environmental aspects

The Standard requires the organization to identify the environmental aspects of its activities, products or services that it can control *and over which it can be expected to have an influence*, in order to determine those that have or can have significant impacts on the environment. (The phrase in italics refers to indirect aspects – which is expanded upon later.)

This is without doubt the most important part of the Standard. All the other elements are linked to this fundamental concept. It is the area where the implementing organization must spend the most time. Indeed, it is also the area where the certification body spend a significant amount of time examining and auditing.

Thus for each individual organization, the possibly long list of environmental aspects identified during the preparatory environmental review will reduce to a more focused list of the most highly significant environmental impacts.

How environmental aspects are identified
Much of what follows will have been performed during the PER. However, the PER is generally a 'one off' audit and environmental aspects identification must be a dynamic process. The approach can be used of course for any new product, process or service that the organization is considering but the organization must first understand what aspects of its activities may cause adverse change. The main causes of change to the environment fall into the following groups:

- Use of resources such as energy, raw materials

- Releases into the atmosphere from normal activities – dust, noise, heat, odour, waste

- Accidental releases to the environment – such as fires (smoke and toxic gases) and leakage of chemicals

- Development of land, including visual impacts, landscaping, drainage, pest control, changes to natural habitats

- Products and by-products

Having identified what activities may cause changes to the environment, it is then possible to determine what those changes will be.

As an example, a plastics company, using a wide variety of plastics such as PVC, PET, polyesters, polyurethanes and nylons should gather inputs from all its internal experts (personnel who have a wealth of experience of their area of activity within the organization). A good area to start would be with raw materials – plastic powders and granules – because their consumption will be the highest in the manufacturing process. Each plastic type will have associated environmental aspects which can be identified by asking the following questions:

1   Is it derived from a non-renewable resource?

2   Does it require a high input of energy (heat) for it to melt and be moulded into a finished article?

3   At the end of its life, is it capable of being easily and economically recycled?

4   Does it have undesirable by-products during moulding (gases, particulates etc.)? There is also a wide range of ancillary raw materials which are used in the creation of the finished articles:

– Organic plasticizers	– Anti-oxidants
– Stabilizers	– UV stabilizers
– Processing lubricants	– Solvents
– Fillers	– Inks
– Pigments	– Packaging

Each one has its own individual environmental aspects which must be evaluated for significance.

The above process requires a methodological approach with figures for usage, scrap, waste sent to landfill etc. (obtained wherever possible) to enable some level of objectivity when making evaluation for significance.

It might be difficult for the environmental management representative to grasp the details here for each department and so one way of gathering information would be to get each area to complete a questionnaire (as suggested previously in this chapter when discussing performing the PER). Certain individuals will have a wealth of knowledge because they have been employed in the organization for a number of years. Although some of this information may be informal and perhaps subjectively based, if it is analysed a broad picture will emerge.

For example, over a period of years the business will have been subjected to the requirements of regulations and legislation. There will be personnel within the organization who will possess knowledge that is relevant to particular legal obligations. Such knowledge may of course be superficial. An employee may have been responsible, over a period of time, for completing the necessary legal documentation related to solid waste being sent to a landfill site. Nevertheless, this experience is valuable and should be utilized by the organization. Clearly, the message is to use the existing knowledge within the organization.

How environmental aspects are evaluated

The Standard places much emphasis on the word 'significant' and the judgement of 'significance' is a critical issue, which bears upon a fundamental conflict between, on the one hand, the need to ensure that important aspects are not overlooked by cursory assessment and, on the other hand, the need to pay attention and assign resources (in a responsible manner) to those aspects which are truly important. The difficulty is exacerbated by the absence of any universal measure for comparative assessment of widely different environmental impacts.

Because the Standard cannot give any detailed guidance on this recognised difficulty – it is a generic standard – Sector Application Guides (SAGs) were developed by trade associations to assist members in this area. A list of trade associations that have developed SAGs in the UK can be found in Appendix II.

For example, some SAGs may advise on certain technologies or practices to be avoided – or certain raw materials. One SAG, written for the chemical industry, advises that emergency response procedures should include media control. During major chemical spillages this takes on a major significance. Textile organizations are advised to change from bleaching processes using chlorine to less environmentally harmful, peroxide process.

No definitive style or approach is in evidence throughout the SAGS (having being developed by different writers and committees) and this may be an area requiring control and standardization in the future. There are no rights and wrongs for aspects evaluation and a number of approaches have been suggested by environmental consultants; however, the soundest, and most widely used by implementing organizations, is based upon consideration of risk assessment.

The tool of risk assessment

Risk assessment, which has been a management tool for many years (especially in health and safety issues), has two components:

1   The likelihood (or probability) of an incident

2   The likely consequences of that incident (or gravity of the incident)

Using a simple analogy, consideration of the 'risks' involved when travelling by commercial airliner will tell us that the likelihood of an incident (accident) occurring is very low, but the consequences of that incident are severe (often fatal).

By comparison, the risks of having an accident when travelling by car show us that the likelihood of an accident is higher, but the consequences are usually minor or trivial (broken limbs, cuts and bruising, scratches, shock etc.)

So, if the probability of an incident is low (because of procedures within a management system – be it health and safety, or environment – and/or technology) but the consequences are high (pollution, prosecution, loss of image and market share) then the risk is still an item to be considered as having some significance.

There is some fairly sophisticated software available to assist in aspects evaluation; these computer programs have their uses but they invariably rely on some subjective decision-making by the user. A basic approach that can assist the process of evaluating risk assessment is to construct a simple matrix:

Likelihood of occurrence		Consequences of that occurrence		Significance
low	x	low	=	insignificant
high	x	low	=	moderately significant
low	x	high	=	significant
high	x	high	=	highly significant

Thus only the product of 'high x high' will score as highly significant. This matrix has its merits initially but the organization should develop a more sophisticated model. If not using a software program, then a 'manual' example follows.

A table can be drawn up with the environmental aspects of the organization in the vertical column, and the criteria against which the aspects are assessed along the horizontal axis (see Figure 3.3). It is at this stage that a simple scoring system needs to be used. The advantage of using a scoring system is that numbers, and the products of those numbers (or even simple addition of those numbers) are easy to follow. It is easier to form a picture of what is being achieved using numbers than it would be using words. It is only a mechanism for allowing us to simplify a complex undertaking. So instead of having an order of significance starting at 'extremely significant' and going to 'highly significant' then 'moderately significant' and ending up with 'insignificant', a system of numbers between 1 to 10 (with 10 being the highest significance) is far easier to manage and understand.

The system uses a key of:

1	low significance
2	medium significance
3	high significance

REGISTER OF ENVIRONMENTAL ASPECTS

Issue	Risk (1) Probability	Consequence	Product of risk	(2) Past Incidents	(3) Nuisance	(4) Abnormal	(5) Local/Regional/Global	(6) Time Scale	(7) Future Activities	(8) Legislative or other obligation	(9) Lack of information	Score of significance	Discussion of Significance
Emissions to air	3	3	9	2	2	2	2	2	1	3		23	*Awaiting new precipitator*
Emissions to water	2	3	6	2	1	2	1	1	1	3		18	
Solid Waste	1	1	1	1	3	3	1	2	1	2	3	17	*Require weights of solid waste*
Other Waste	2	1	2	1	1	1	1	1	1	1	1	10	
Use of Energy	1	1	1	1	1	1	1	1	1			7	
Noise	1	2	2	2	1	1	1	1	1	1	1	10	
Visual Impact	1	1	1	1	3	2	1	1	1	1		12	*Several complaints from neighbours*
Ecosystems	1	2	2	1	1	1	2	1	2			10	
Transport	1	1	1	1	1	1	1	1	1	1		10	
Suppliers	1	1	1	1	1		1	1	1	1	3	10	*Questionnaires being sent to suppliers*

Figure 3.3 Register of environmental aspects

The following nine criteria are cross-referenced in the table:

1   Risk to the environment. In terms of the nature of the hazard, the probability of its occurrence and the likely consequences, should an incident occur?

2   Past incidents. The numerical scoring system allows more scope for looking at the history of operation of a process. For example, if the process is thought to be well controlled but there have been problems of an environmental nature in the past, then it could be argued that the likelihood of an incident is higher compared with the incident-free record of other processes. The site of the organization may have been contaminated by previous use and may always have the potential for a pollution incident.

3   Actual or potential nuisance. Is there any nuisance to neighbours – monitored, perhaps, by complaints? Is there any potential nuisance from a proposed expansion to the site?

4   Significance and process conditions. Significance should also be considered in the context of normal, abnormal (start-up/close-down periods) and emergency working conditions. Some processes have a much greater potential for pollution or even an environmental incident during start-up than when the processes are stabilized.

5   Spatial scale. Another way of categorizing environmental impacts is according to their spatial scale. Impacts, both direct and indirect, may be divided into local, regional, national and global. *Local* effects are those occurring within the local vicinity of the cause. *Regional* effects are those which extend beyond the locality of the cause but are still spatially limited. *National* effects are clearly those which occur on a national scale and *global* effects are those occurring on an international scale.

6   Time-scale. An additional categorization method could be to consider the time-scale over which environmental impacts occur. Occasional impacts which have no acute, or long-term, consequences may be viewed as less serious than effects which are likely to occur frequently and which have serious and/or long-term consequences.

7   Future activities. What strategic plans are in hand for expansion of the business or for producing a new product line or production process?

8   Legislative requirement. If an organization's process is subject to legislation (a prescribed process) or discharge consents, then it is argued

that, by definition, these environmental aspects are significant – otherwise why would such legislation be in force? A third party is, in effect, telling the organization that the issues are significant.

The consequences of infringement could be severe: heavy fines, imprisonment for company officers, negative publicity etc.

9   Information. If there is a lack of information on which to make a satisfactory appraisal, then this issue should automatically become a significant environmental impact, and stay at that status until further information proves it to be otherwise.

Points to consider during the evaluation process
If environmental impacts cannot be evaluated by direct measurement methods (such as those used to measure waste streams; volume and rate of discharge; heavy metals; acidity or alkalinity, etc.) it is recommended that calculations involving mass balances (inputs versus outputs of a process) or computer modelling are utilized. Provided that the basis of such use is sound, no external auditor would make this an issue.

Care should be taken when using numeric systems, however, that the numbers derived make sense. An over-complex system may give a possible scoring range of 0-300, which is very unwieldy. If such a scoring system only generates numbers in the range 10-50 then there is no point in having such a wide range. A scoring range of 1-10 has been found to be adequate in practice: low significance receiving a score of 1 and high significance equalling 10, with the majority of 'moderately significant environmental impacts' clustered around 4-8.

Once the evaluation process has been performed, it makes sense to stand back and look at the numbers. For example, an organization may arrive at a range of numbers from its matrix, from 0 (insignificant) up to 100 (highly significant). However, the majority of scores may well cluster around 50 or 60 with differences of only 1 or 2 between. Given the amount of subjectivity involved, it may well be that choosing an impact with a score of 59 as being more significant than one with a score of 58 is an error and that the level of significance could in fact be reversed.

As with any scoring system, common sense must prevail and in areas which are 'grey' some subjectivity must be used. Provided that such subjectivity is equally applied throughout, then this will withstand scrutiny by the certification body.

Other organizations use a combination of letters and numbers to create an alphanumeric score. However, it is important to remember that the scoring system is not absolute – it is merely a means for the organization to make sense of a very complex set of environmental concepts and interactions. Even with these simple sets of numbers, some subjectivity is inherent and the final numbers should be examined and, perhaps, an element of 'weighting' put into the numbers themselves. This weighting may arise from circumstances peculiar to the individual organization. For example, proximity to a National Park boundary may give rise to a higher number of environmental complaints than would be 'normal' for this particular industry sector.

Each implementing organization will develop its own model. Provided that the reasoning behind the process is sound, the certification body will not find fault with it.

Life cycle analysis

Implementing organizations stumble sometimes in deciding how much of the life cycle of a product should they look when assigning significance to environmental aspects. There is no strict requirement in the Standard to perform a 'life cycle analysis' but if an organization has the resources to do this it can be beneficial. (See Appendix I for a definition of life cycle analysis.) A life cycle analyis may assist in future design changes and will demonstrate to the certification body an extra awareness of wider environmental issues. Standards are becoming available in this important area (see Appendix III).

Some organizations direct much effort into minimizing the future environmental significance of a product – by virtue of the design of the product – at the early stage of the life cycle. They argue that because of this, the effects on the environment during use by the purchaser, and the impacts on the environment at the end of the product's useful life, are minimal and not significant.

Environmental impacts can be categorized into two groups – direct impacts and indirect impacts. A *direct impact* is a change arising as a direct result of an activity under the control of the organization. An *indirect impact* is a change that arises as a result of someone else's activities; these activities are connected to the organization in some way but are less easily controlled as they can only be influenced indirectly.

Direct impacts

Direct impacts are the easiest to consider. Indeed, the majority of organizations will commence their environmental aspects identification in this area before considering indirect aspects. There is nothing wrong with this approach provided that evaluation of the aspects does not begin until all the indirect aspects are known. The reason behind this is that an indirect aspect may rank as being a highly significant impact, far outweighing any of the direct impacts of the organization.

Direct impacts are usually far easier to measure and monitor than indirect impacts and have been addressed in some detail in preceding chapters. The above matrix focused mainly on direct impacts.

Indirect impacts

Identification and assessment of significance of indirect impacts represents an altogether more difficult exercise. Although subjectivity was bound to be associated with direct impacts evaluation, there is even more scope for subjectivity when evaluating indirect impacts.

Although Chapter 2 only mentioned indirect aspects in terms of supplier and customer environmental behaviour the organization should also bear in mind some of the wider indirect environmental issues. Some of these issues lead to contradictions in terms of what an organization should do to demonstrate 'green' behaviour:

- Resource depletion versus lifestyle considerations – manufacturing packaging for gift-wrapping adds little or nothing to improvements in the quality of life.

- Global warming – should an organization consider not using any fossil fuels and rely on manual labour for maufacturing?

- Product life cycle considerations – should all products be manufactured with built-in obsolescence: good for continuing business reasons but wasteful of resources!

- Recycling and waste management – should all products be designed to be 100% recyclable or with minimal waste management requirements? The answer is yes, but the research and development costs would not make this commercially viable in a smaller organization or one with low-value products.

Some indirect aspects can almost be 'hidden away' in an organization's internal management policies. An example of a hidden issue is that of an organization that has a company policy of free fuel as a 'perk' to some of its managers. This is not really conducive to good environmental management if it transpires that employees ignore the distance they travel to work every day and forego using public transport in preference to their cars. Here, indirect impacts are being created by the organization's policy. Encouraging use of public transport through ticket discounts, and so forth, could be a better policy option. In addition, moving from an old inner-city location to an out-of-town industrial or business park with purpose-built offices will probably save money on heating and lighting costs but with no public transport to such a location, will force existing employees to use cars, thereby adding to traffic congestion and air pollution.

Changing such 'perks', of course, could lead to highly-emotive arguments between managers and the organization's policy makers.

However, the intention of the standard is to provide a focus on the most appropriate way forward. In practical terms this means only a consideration of those indirect environmental impacts closest to the organization – the impacts of suppliers and customers.

Also, organizations want to know how far up and down the supplier and customer chain they should look for environmental probity. The life cycle analysis approach is to consider all the environmental impacts a product has, from its 'cradle to the grave'. However, as stated, this is not specifically required by the Standard. Thus the focus here will be on practical steps to take with suppliers and customers.

Approach to the supplier chain

At the early stages of supplier probity evaluation, it would not be practical, prudent or meaningful to request 'questionnaires' to be completed by all suppliers. This seems to be a carry-over from quality assurance philosophy. Certainly this gives the appearance of a 'doing something useful' exercise. However, it is better to initiate a management programme with the objective of gaining sufficient knowledge of the suppliers' businesses to structure a programme accordingly (setting out objectives and targets). Such a programme could have a set of objectives as below:

Educate:
Rather than use a heavy-handed approach, it makes sense to educate suppliers on environmental issues. It may well be that they are totally

unaware that some of their activities have an environmental impact. Educate by visits, discussions and even some practical assistance.

Develop a criteria:
Develop a set of measures that will be used to categorize suppliers.

Categorize – inform suppliers:
Having evaluated suppliers' probity it makes sense to tell them what rating they have achieved. Such ratings should be simple and informative. (For example, the supplier should endeavour to purchase only timber from sustainable and managed forests to improve its current rating.)

Develop a purchasing policy:
The realities of purchasing must be taken into account. If there is a sole supplier of a raw material, whose environmental policies are in question, then it is somewhat nonsensical to subject this supplier to filling in questionnaires when the organization is going to have to purchase from them anyway. Far better to start a management programme with the objective of sourcing a more environmentally responsible supplier. This may take some time but it can be a long-term objective within the environmental management structure.

The following is illustrative of the above approach:

- A *chemicals manufacturer* is likely to exhibit high direct environmental impacts of both a polluting and a resource-usage nature. This organization will therefore have highly significant direct impacts which should be addressed first, before considering the indirect aspects of suppliers.

- A *chemicals stockist* will have some direct environmental impacts but of more significance will be the chemical suppliers (for example, indirect impacts). Stocked items need to be transported both to the stockist and the customer. Sub-contracted delivery services are an indirect environmental issue of significance. (The transport of an organization's products is sometimes overlooked. If the transport is sub-contracted then this is a purchased service from a supplier and this supplier is no different from any other 'materials' supplier.)

- A *plastics packaging manufacturer* will have some direct environmental impacts, such as workplace atmospheric pollution, workplace and neighbourhood noise, pollution and high electrical energy usage. However, there will also be some significant indirect impacts. These indirect impacts arise because the plastics supplier will be using non-renewable resources (plastics and energy derived from mineral oil) and the end-user, the customer, will be disposing of the plastics packaging. As packaging film is, by its very nature, destined for a short life followed by disposal, the indirect impact is highly significant.

Another step an implementing organization can take is to measure the level of awareness of environmental issues achieved by its suppliers. This information can be gleaned from very careful scrutiny of the answers obtained from questionnaires. Questionnaires were previously said to be an inappropriate mechanism for 'testing' the environmental probity of *all* suppliers. However, they have their merits if used with discretion – aimed only at major suppliers.

By careful phrasing of the questions in these questionnaires, an insight into a supplier's environmental awareness will be gained. The objective could then be to raise awareness by visits, offers of advice, or the inclusion of suppliers in your internal audit schedules. An organization should be able to measure awareness increase by sending a second questionnaire – perhaps 12 to 18 months later – and compare these replies with the previous answers. (See Figures 3.4, 3.4a, 3.5 and 3.5a for a fictitious example – based on an amalgamation of several organizations – that shows such an increased awareness merely by comparison of the answers.)

Some organizations' purchasing policies may be dictated to them by the corporate body. This body may well be located overseas and operate to a different agenda or timetable of global environmental purchasing policy. The corporate body may well dictate that only named suppliers are to be used to gain the advantage of lower prices by purchasing in bulk.

This may conflict with the local purchasing policy of the ISO 14001 implementing site and is obviously a cause of concern. The way forward here would be to establish an objective to influence the corporate body, perhaps to gather evidence that the named suppliers are acting in a way that does not demonstrate environmental responsibility.

Environmental Performance Questionnaire

Company Name: *Clean Image Company Ltd*

Address: *Newtown*

Green Acres

Devon UK

Representative on Environmental Issues: *J. Widget   QC Manager*

Products Supplied: *Adhesives, solvents, mastics ancillary equipmet*

If you have been certified to an Environmental Standard, please enclose a copy of the certificate. The rest of the questions do not need an answer.

What do you believe to be the environmental impacts of your business activities:	*We do not pollute anything*
Do you carry out regular environmental audits:	*Site inspections*
Do you have a written environmental policy:	*Yes - written by our sales manager*
Does your organization operate any processes which require authorization from National Enforcement Authorities:	*?*
Do you require consents to discharge effluent into local rivers or sewers:	*?*
Have you breached any of these authorizations or consents over the past 5 years:	*Never had any problems with the factory inspectors*

Figure 3.4 Environmental Performance Questionnaire (p1)

Environmental Performance Questionnaire

(continued)

Do you attempt to recycle any materials or products?	*Yes. Plastic containers re-used. 50% of scrap recycled into black mouldings.*
Do you check out the environmental performance of your suppliers? How do you do this?	*Yes. But only with 5 potentially significant suppliers. Initially sent questionnaires followed up by visits and dialogue.*
Do you carry out any environmental initiatives?	*Member of local Agenda 21.* *Assisting local college with a recycling project.*
Do you intend to work towards obtaining a recognized environmental management standard? If so which one?	*Aiming towards ISO 14001 by next year. EMAS is not being considered as our markets are mainly outside Europe.*

Completed by: *J. Widget*

Position: *QA Manager and Environmental Co-ordinator*

Sign: *J Widget*

Date: *21/3/97*

Figure 3.5a 2nd Environmental Performance Questionnaire(p2)

Approach to the customer chain

The requirement of the Standard for an implementing organization to influence the environmental behaviour of their customers, is without doubt, a difficult requirement to meet. After all, an organization cannot dictate how a customer should behave – they will lose that customer! Some answers are offered below. The amount of work necessary can be reduced by considering the concept of significance. An organization should establish from its customer database which customers are the largest in terms of volume of product they take, and which customers take products that have the most potential to impact on the environment. These customers will then be the first to be targeted by the organization, when managing its indirect environmental impacts.

The question of how far influence should be exercised along the customer chain can be answered in terms of significance. Look no further at this stage than to your distributor (for example). For an organization supplying brake shoes for vehicles, it would be totally impractical if it attempted to influence every end user of its brake shoes (that is, the general motoring public). Indeed how could it check whether such influence was working? A more practical approach is to attempt to influence either corporate bodies such as local authorities (who will purchase brake shoes in large numbers for their fleets of lorries, buses, snow ploughs, etc.) or the larger distributors who distribute to garages and spare parts stockists. By influencing a few distributors, a significant number of used brake shoes could be recycled or disposed of in an environmentally responsible way.

What form can the influence take?

One practical and relatively low-cost approach is to collate, review and summarize existing published information on one or two of the most significant indirect impacts.

- Continuing with the example above, the organization might fund research into alternative ways of producing brake shoes without use of asbestos.

- For an organization printing telephone directories (which have a limited life due to the constant number of changes) a financial incentive to return obsolete directories to a central collection point for recycling could be implemented.

- The manufacturer of plastic packaging could state on the packaging itself what type of plastic it should be recycled with. This may encourage some customers to demonstrate environmental responsibility.

In conclusion, it should never be the objective of an organization to make all suppliers and customers obtain ISO 14001. For some smaller suppliers in particular, the task of implementation may be very costly and achieve very little – their environmental impacts may not be appropriate or significant when compared with larger suppliers. Only the suppliers and customers who have a significant impact should be targeted for ultimate compliance with ISO 14001 – if this is thought to be the best way forward. Certainly, the history of quality assurance has shown us that smaller organizations have struggled to comply with the requirements of the Quality Standard ISO 9000 and have designed inappropriate systems. It was never intended that ISO 9000 compliance be 'forced' upon smaller organizations – who may have no practical need for such a system. The same is true for ISO 14001.

At all times, the implementing organization should ask itself: 'What are we trying to achieve – bureaucratic paper systems or a real contribution to reducing adverse impact on the environment?'

Clause 4.3.2: Legal and other requirements

The Standard requires that organizations *shall establish and maintain a procedure to identify and have access to legal and other requirements to which the organization subscribes, that are applicable to the environmental aspects of its activities, products or services.*

Examining the 'other requirements' first, these might include a situation where, for example, the corporate headquarters of an organization decrees that certain solvents will be banned from all sites by a certain date. Clearly this directive must be obeyed and reflected within the overall environmental policy, programme, objectives and targets. As a further example, the use of the plastic PVC is under active environmental scrutiny because it is argued that, as PVC is a chlorinated material, it can have adverse health effects on humans and wildlife. Although such evidence is inconclusive, an environmentally responsible organization may mandate that all of its sites phase this material out until there is further research into the environmental impact and safety hazards of PVC.

Similar codes of practice will operate within other industry sectors. Local by-laws may also be applicable and these can be obtained from local authorities or water companies. Most countries have government environmental agencies who will assist with national legislation information.

The legal requirements will, of course, be considered by many to be the most important due to the possibility of fines, imprisonment or the forced closure of the business. Therefore, keeping up to date with legislation is important. However, bearing in mind the extent of the environmental legislation emanating from regulatory bodies throughout the world, this is no simple task.

For most organizations, it will be necessary to rely upon assistance from either their environmental consultants or the specialists who collate and publish details of legislation. These tend to be trade associations, law firms or specialist publishing companies. Some sources (most applicable in the UK only) are listed in Appendix II. These sources will supply information of a general nature which will be relevant to the organization's activities.

It is highly likely that in some organizations, legislative knowledge is already possessed by certain individuals. For example, the transport department may well have records of waste sent to landfill and the necessary legally required paperwork and regulations to hand. The production manager may well know the requirements regarding discharging only certain consented amounts of process water to sewer. However, such knowledge may be fragmented and it therefore makes sense to collate information concerning applicable legislation in the form of a file or a register. A typical list could include legislation pertaining to:

- Land and buildings
- Raw materials
- Air emissions
- Water discharges
- Waste management

- Chemical usage – and storage
- Working conditions
- Processes
- Products
- Transport

Clause 4.3.3: Objectives and targets

The Standard requires that *the organization shall establish and maintain documented environmental objectives and targets at each relevant function and level within the organization.*

Objectives should be the longer term goals derived naturally from the environmental policy. It should be understood that each identified significant aspect will have an associated objective. Quantification can then take place through monitoring and measurement in order to meet such goals. Of course, all objectives and targets should be realistic (with rational decision-making behind them).

Objectives should be fairly specific; every significant environmental impact must have an objective, or objectives, set against it. Following on from this, each objective should have a measurable target to demonstrate that the objective is being attained (or otherwise). Targets are more specific, more easily measurable detailed performance requirements which evolve from the objectives and allow an organization to monitor whether the stated objectives will be achieved. Early warning mechanisms for targets not being met should be in place – via the process of regular reviews and audits.

Quantification of targets is not always easy but sometimes there are suitable measures within the organization's existing management systems.

A broad objective could be to reduce waste by 10% compared to the previous year. One of the targets could be quite specific: to reduce waste to landfill. Deciding how to measure this could cause a problem until it is remembered that records will exist within the organization (from duty-of-care notes, weighbridge tickets and also from invoices for cost of removal, cost of landfill). It may only be a matter of obtaining this information and manipulating it to achieve the measurements required.

How to quantify targets

Some targets will be dictated by the requirements of directives or legislation and therefore are decided upon outside the organization. That apart, using the example of landfill waste above, if an organization has identified from its preparatory environmental review that solid waste to landfill is a significant impact, what should be the target to aim for to reduce this amount of waste?

First and foremost, quantification of what is actually sent to landfill needs to be obtained. As stated above this could probably be obtained from weigh-bridge tickets and other records. If, from these records, it is discovered that in the previous year 100 tonnes were sent to landfill, how does the organization derive a meaningful figure for reduction? Using percentage figures, is 1%, 10% or 50% the correct figure?

On examination, an improvement of 1% is meaningless as far as environmental significance is concerned as it may be difficult to measure. There is also the fact that the costs of the controls for this small reduction may outweigh any financial considerations – always an issue in any organization. The improvement of 50% would appear at first glance to be commendable but on closer examination is probably somewhat unrealistic. Otherwise, why has the organization not done something about such a large improvement as this before?

Thus 10% appears to be a starting point and an achievable target – with measurable associated cost savings. If this proves to be slightly off the mark, then it can be modified as the management system matures.

However, of the 100 tonnes of waste being considered, a further break-down into categories or actual sources of waste shows:

- The administration facility generated 5 tonnes of paper waste.

- The production facility generated 95 tonnes of mixed solid waste.

Therefore, targeting the administration facility as an area of improvement in waste figures would only show a small reduction. The waste produced by the production facility should be focused on as the area where a significant improvement can be made.

Analysis of the production facility can be broken down further:

Production facility	Tonnes of waste produced
Folding department	50
Reaming department	1
Slitting department	10
Printing department	34
Total	**95**

Therefore, focusing on savings in both the folding and printing departments will save the largest percentage of waste and have the biggest impact in achieving the targets the organization has set itself. All the above can be outlined in a management programme, with operational controls to segregate, measure and reduce waste in a focused manner (using the limited resources available).

Obviously, only running the factory at 50% capacity, due to slack customer demand, will reduce waste by a roughly corresponding amount. Unless this is taken into consideration in the calculations, errors in the figures will occur.

Clause 4.3.4: Environmental management programme

The Standard requires that *the organization establish and maintain a programme, or programmes, for achieving its objectives and targets. It shall include:*

a)   *designation of responsibility for achieving objectives and targets at each relevant function and level of the organization;*

b)   *the means and time-frame by which they are to be achieved.*

Such a programme, or programmes, will show whether there are adequate resources – personnel, plant and equipment, and finance – to achieve the stated objectives. Additionally, they will show if extra training is required.

The management programme(s) should also address new or planned activities as well. They should show considerations of planning, design, production, marketing and disposal stages.

This must all be planned at an executive level and a typical logical sequence for such planning can be drawn up in the form of a chart (see Figures 3.6 and 3.6a).

The chart shows the linkages between the total number of identified environmental aspects, both direct and indirect, and those that are significant. The linkage is then shown between those significant impacts and the resultant objectives and targets. It should be noted that there can sometimes be a number of objectives to achieve only one target.

Note that the example programme assigns responsibility to several individuals and not just the environmental manager. (This is true management responsibility: distributing responsibilities across the organization.) Also, other programmes could co-exist to cover shorter or longer time scales.

Management Programme: -1997

Environmental Aspects	Significant impacts	Objectives	Targets	Time Scale	Responsibility
Solid waste - cardboard, paper, filter cake	Filter cake - contains heavy metals	Modify process to reduce heavy metals	Cd and Zn levels to 5%	First 1/4 of the year	Production Manager and heavy metals team
Use of energy - gas, electricity and oil	Possible oil leakage from oil storage	To reduce reliance on oil / Investigate costs of gas heating	Report by 2nd 1/4 / New bund wall		Facilities manager
Visual / Noise from compressors / Dust from blowing down / Smell from plasticisers	Complaints from neighbours	External site housekeeping inspections	Every month as part of compliance audits	Environmental senior auditor	Environmental Manager
Water useage	Use of town's water	Recycle / Test for borehole	10%	By end of 1997	Environmental manager
Old storage tanks	Leakage of diesel / Fire risk	Remove		End of 1998	Estates manager
Effect on aquatic-based organisms by discharge of acidic heavy metal mixtures		Review integrity of all pipework for rinse water and culverts	1) Obtain underground maps 2) Identify leaks	Sept 1997	Laboratory Manager

Figure 3.6 Management Programme - direct aspects

Management Programme: -1997

Environmental Aspects - indirect	Significant impacts	Objectives	Targets	Time Scale	Responsiblity
Investigate suppliers' environmental probity	Potential impacts from poorly managed raw materials usage	Obtain information Derive a purchasing policy.	Top 3 main suppliers in Q1 Next 5 suppliers by Q3	Throughout 1998	Purchasing Manager
Sub-contractors used for storage and distribution of products	Heating of warehouses; fuel use during distribution; packaging used for products.	To audit warehouse facility for environmental probity Formulate policy for all sub-contractors	Target of 2 largest warehouses by end of Q2	During 1998	Environmental Manager
End of life of products by customers	Loss of precious metals to society unless recycled	To devise an incentive scheme for customers to return for recycling	Pilot a questionnaire to customers (10%) to see if a viable scheme by Q4 1998	1998-1999	Sales Director

Figure 3.6a Management Programme - indirect aspects

Clause 4.4: Implementation and operation

The Standard requires, via the following seven sub-clauses, the organization to put into place controls over all activities which have, or may have, a significant environmental impact. Procedures will need to be written and records kept of measuring and monitoring activities.

Clause 4.4.1: Structure and responsibility

The Standard requires that *roles, responsibilities and authority be defined, documented and communicated in order to facilitate effective environmental management*. The successful implementation of an environmental management system calls for commitment from all employees in the organization.

The purpose of this sub-clause is to ensure that personnel are assigned responsibilities for part of the environmental management system and have a clear-cut reporting structure. Job descriptions, or project responsibilities from the management program, may cover this requirement.

A *management representative* needs to be appointed. This can be (and is for the majority of companies) an existing member of staff who, regardless of other duties, has responsibilities for co-ordinating the activities of the environmental management system. There needs to be a direct authority linkage. For example, in the case of a potential environmental problem, the line of communication to senior management needs to be short so that action can occur readily. Commitment begins, of course, at the top level of management, but it is accepted that in larger organizations responsibility to be the management representative is often delegated to a less senior individual . However, in smaller organizations, the management representative might be the managing director himself.

Some organizations spend much of management time defining job responsibilities via documented job descriptions. Indeed, these job descriptions are an excellent method of addressing this clause of the Standard. However, care should be taken to describe the authority that an individual has in an emergency environmental situation. For example, an operator at a remote location may not have the authority to turn off a part of a production plant to prevent a spillage in an emergency situation (this would cause financial loss to the organization due to lost production). The operator may therefore need to refer to a higher authority for a decision which, of course, could lead to the incident becoming more serious due to the time delay.

It may well be that after the event, the operator could have made that decision and would have had the full support of management in dealing with the consequences of lost production (such as lost revenue plus, perhaps, the additional cost of nonconforming product requiring disposal). Thus such levels of environmental authorization should be defined in job descriptions so that all individuals are aware of what decisions they can make, especially in an emergency situation.

Clause 4.4.2: Training, awareness and competence

The Standard requires that *the organization shall identify training needs and this clause requires that all personnel whose work may create a significant impact upon the environment have received appropriate training*.

Accreditation criteria indicate that, prior to certification, an organization must show that all key staff (that is, those involved in managing significant impacts) have undergone a training-needs analysis and have received training accordingly (see Annex IV).

Thus the organization must satisfy the following four criteria:

1   **Ensure that training needs are identified:**

This can be performed via appraisals. In most companies this is an annual event – at the very least for salary review purposes. From this, training needs will be identified and a plan of either internal or external training planned. All individuals will need some level of training in the requirements of the environmental policy and a background to the requirements of ISO 14001. Some individuals will need specific training in emergency response. Others may need their roles to be changed and defined. An internal quality assurance auditor may well need to be 'converted' to an environmental systems auditor via an external training course.

2   **Ensure that these planned needs are met:**

There must be a system to ensure that such individual training plans are carried out as intended. Procedures will be needed to describe such mechanisms, as well as including a broader description of how the organization's training strategy is structured. In addition to specific external courses or seminars, internal workshops and briefings are an acceptable vehicle for training. Internal environmental newsletters are also part of the range of tools available.

3   **Verify that the training has achieved its purpose – increased awareness:**

This verification can be performed via feedback from training sessions: either a written report from the individual or a simple questionnaire to complete. Some organizations will ask personnel to undertake simple 'exams' to measure the effectiveness of the training.

4   **Verify that following training, the individual is competent at applying the awareness gained to their particular job:**

This can be achieved by monitoring an individual's work, noting any improvements in work or, conversely, monitoring any persistent failure to absorb such training (for example, by not being aware of the consequences of departure from a specific work instruction).

(The internal environmental auditor, mentioned in paragraph 1 above, can be 'tested' for competence against the appropriate standard: for example ISO 14001 – see Appendix III.)

Contractors, working on behalf of the organization, must also be subject to training requirements and this must be addressed in the training procedures. The employees of the contractor should have a certain level of training – that level to be determined by the organization. For both contractors and the organization itself, it must be kept in mind that the concept of significance must be applied to any training plans or programmes. The individual who is in an environmental front-line position – whose actions have the potential to cause a major impact on the environment – should have priority in environmental training followed by a more intensive scrutiny of awareness and competence than the individual whose actions have little potential to impact on the environment.

Motivation of personnel is sometimes overlooked. It must never be forgotten that personnel, even if well-trained, need to be highly motivated to ensure successful implementation of the environmental management system.

Clause 4.4.3: Communication

This sub-clause refers to all types of communications, both within, and external to, the organization. It requires organizations to *establish and maintain procedures for internal communications between the various levels and functions of the organization*. It also requires organizations to document and respond *to relevant communications from external interested parties*.

Examples of *internal communications* are:

- Communicating environmental objectives and targets to employees

- Raising awareness of environmental issues to employees

- Communicating the environmental policy to employees

- Advising of nonconformances to relevant departmental heads

- Reporting incidents arising from abnormal or emergency operation to senior management

Examples of *external communications* include:

- Dealing with complaints or writing pro-actively to schools and colleges offering student visits to the site for educational purposes.

- Dealing with the media, especially in the event of an incident.

An inability to communicate effectively within the first few hours of an incident will seriously reduce the company's ability to control the situation. This will undermine the company's reputation in the minds of staff, customers, the media and the public. (See also sub-clause 4.4.7 *Emergency preparedness and response*, later in this chapter.)

Clause 4.4.4: Environmental management systems documentation

The organization shall establish and maintain information, in paper or electronic form, to:

a) *describe the core elements of the management system and their interaction;*

b) *provide direction to related documentation.*

It is not the intention of this book to describe in detail how to write manuals and procedures to meet the requirements of the implementing organization and the Standard. In fact, there is no laid down format and each organization should develop its own style that it can work with. However, there are, within industry and commerce, certain styles (based upon years of experience) which are better than others. Meeting the needs of the organization and complying with the Standard is the first consideration; being open to audit is a close second.

The following briefly outlines a practical documented system structure but leaves the detailed style of the environmental manual, procedures, work instructions etc. to the implementing organization. It is certainly a good idea to have a three- to four-level 'pyramid' type hierarchy of documentation, with the environmental policy at the top, which spreads sideways down through the manuals, procedures and supporting documentation. This is the structure to be found in the majority of organizations that have a management system but, again, there is no one 'correct' structure:

First level:	The environmental manual
Second level:	The procedures
Third level:	Work instructions/specific routines
Fourth level:	Standard forms and documents

This structure is illustrated in Figure 3.7. Each of these levels is described below.

The environmental manual

This can be a rather sparse 'manual', which need only include the environmental policy and a broad description of how the organization has addressed the requirements of the Standard. As a relatively thin document, with no commercially sensitive information, it can be sent to customers or other interested parties (at little cost) and is, in fact, a good marketing aid.

The opportunity should be taken to show how, for each clause of the Standard, the organization has procedures in place and decision-making processes. These should be referenced. This has the double benefit of showing interested parties (including external auditors) exactly how the Standard is addressed and, more importantly, making this clear for the organization itself. Obviously, when compiling the environmental manual, if the organization cannot assign a procedure or a methodology of working to one of the clauses of the Standard then this must demonstrate a gap in the environmental management system.

Environmental Management System
Documentation Structure

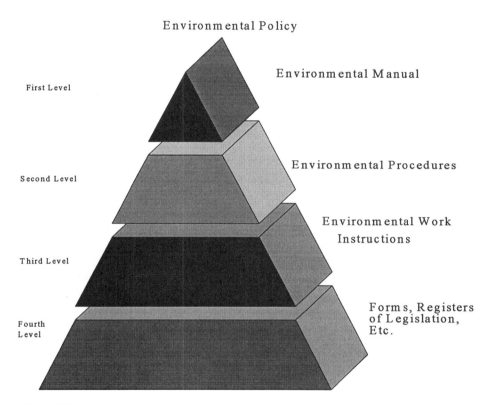

Figure 3.7 Documentation structure

Certainly, the concept of a 'signpost document' is a good analogy and Accreditation Guidance (Appendix IV) refers to a signposting, or 'directory', style for this top-level document.

Environmental management system procedures
The procedures will tend to dominate the bulk of the whole documented system. However, efforts should be made again to signpost, if possible, other existing systems. Thus, if there is a 'handbook' for operating a particular piece of equipment safely, perhaps with environmental considerations as well, then

this handbook need only be referenced. This is preferable to writing a new procedure or instruction when the existing one is adequate.

i) *Identification:*

The identification of a procedure will probably evolve from the management system identifying environmental impacts (ranking for significance, setting objectives and targets to reduce such impacts) and the operational controls necessary to measure, monitor, control and minimize such impacts. All these activities will need documenting in the form of procedures.

ii) *Drafting:*

An individual, following procedure identification, actually has to write the procedure. This is not a daunting task provided there are inputs from personnel who are involved (or will be involved) in operating the procedure. In practical terms, a procedure has to be workable and only those at the 'hands on' level will be able to write such a workable document. This will also give ownership of the procedure at the 'hands-on' level. *Ownership* in this sense means that because the personnel using the procedure actually wrote it, they are far more likely to follow it correctly. This is not always the case if a procedure is written by an external author – perhaps a consultant. Personnel may feel the procedure has been 'thrust' upon them and may be less willing to obey it.

Procedures can be in note form – hand-written at this stage – as it is likely that changes will be required during the pilot phase.

iii) *Piloting:*

Experience shows that it is extremely difficult for an individual to write down their daily tasks; they are of such a second nature that it is easy to make omissions, or to put the steps of a process in the wrong order. Therefore, the procedure should be piloted by other personnel. There may be subtle interfaces or interactions with other personnel or other procedures that the original authors had not considered because they are too close to their own area of responsibility.

iv) Revision:

It is inevitable that revisions will be required. For example: simple errors may need correcting; gaps may have been discovered; a new perspective may have been added.

v) Implementation:

It is important that all those personnel using the procedure understand its purpose. Some time should be invested in training personnel – introducing the reason for the procedure should be the focus of such training.

vi) Auditing:

Last but by no means least, the procedure once written, piloted, revised and implemented needs to be verified as to its effectiveness. This can only be checked by an independent internal audit – and the rule of the '6 Ws' is good guidance to follow. When writing a procedure, if it addresses the following questions, then it is likely to be meaningful and robust:

1) What does the procedure require?

2) When does the procedure need to be followed?

3) Where do the activities take place?

4) Whom does the procedure apply to?

5) Why is the procedure carried out?

6) HoW are the activities carried out?

Using a sense of perspective is also a good guideline. Concentrate on the fundamental reason for the procedure and, initially, 'ignore the spelling' is a good maxim. Spellings can be easily corrected with a software spell-check although most software will not pick up incorrect grammar. Flaws in the logic and purpose of the procedure are not so easily corrected.

Environmental work instructions/specific routines
To begin with, it is not mandatory for the organization to include environmental work instructions in its documentation hierarchy if it is not appropriate to the business. However, in an organization with many processes, comprising of levels of alternative routes or recipes, there may be a need for separate, more

specific and detailed 'routines' for operators. Otherwise, they may find it difficult to locate the relevant section contained within just one long procedure. This could lead to an environmental incident. Therefore, a work instruction at the point of the operation is a good idea. A note above a drainage sink, to remind operators that locally produced acidic waste liquids must not be poured down the sink, is a type of work instruction.

If work instructions become obsolete, and they are followed by an operator, then an environmental incident might occur. Obviously such instructions must be controlled as to revision status and location on the site. (See Section 4.4.5 *Document Control*.)

Standard forms and documents

Every organization uses forms of one kind or another. Forms allow personnel to record information in a structured manner so that other personnel can read and use this information. Forms also remove the requirement for the individual to remember every piece of information given. This means that less errors will occur in the management system as the system is not dependent on the frailty of the human memory. Well-designed forms can act as a prompt for individuals to record the correct quantity of information as well as the quality (that is, the usefulness of that information). Therefore, it makes sense to control the information demanded by the form so that nothing important is overlooked. Forms may be revised many times throughout their life to cope with changing circumstances. Such forms should be controlled, so that if a form is revised this should be indicated on it (Revision 1, 2, etc.)

Of course, the organization should review the 'significance' of the information on the forms and if it is deemed to be trivial, then there is no need to control such a form. It is good practice for any organization to review all the forms it uses, make a list and analyse what function each form performs. Most organizations who perform this task actually end up with less forms in circulation. The remaining forms are more focused in their purpose.

The reference to 'documents' covers all other 'reading matter' that the organization refers to in its environmental operations and decision-making processes. This could include copies of international standards, machine operation manuals, codes of practice, tables of calculations etc. Again, the organization should review which documents are significant (in terms of the information they contain) and, where appropriate, ensure that they are given a controlled status.

These manuals, procedures, work instructions and appropriate forms must ultimately be issued to personnel in a controlled fashion and this is addressed by the next sub-clause.

Clause 4.4.5: Document control

The Standard calls for the organization to *establish and maintain procedures for controlling all documents* required by ISO 14001 to ensure that:

- They can be located.

- They are periodically reviewed.

- The current versions are available at the correct locations.

- Obsolete documents are removed.

- Any obsolete documents retained for legal and other purposes are suitably identified.

Thus, this sub-clause is in effect stating that the environmental management system documentation, as required by clause 4.4.4, should be controlled or managed. The control of documentation can be exercised at the different levels of documentation (as in the three to four level pyramid hierarchy suggested in the previous section discussing clause 4.4.4). For example, stamps can be used to signify 'controlled' or 'uncontrolled'; coloured paper can be used, with or without a special logo, to make it obvious if an 'illegal' photocopy is being used.

Whatever method is used, it should always focus on the principle that the desired and planned information is available to those personnel that require it. There should never be any danger that out-of-date information, or the wrong data, is read or used by those individuals. If information is not important, or it does not matter whether it is up to date or not, then there is no need to control it. Marking it 'uncontrolled – not subject to update' – is a good mechanism to ensure that the casual reader can use that particular document for background information but is also prompted to seek assurance as to its current validity (for example, if a particular decision needs to be made based upon the information in the document).

Controlling documentation does give personnel confidence in their activities; they are sure that the decision they have made is the correct one. Indecision based upon mistrust of information received is responsible for many environmental incidents.

Signatures of the management representative (or even the Managing Director) on each page are also examples of control being exercised in documentation distribution. Consideration should also be given as to the cost-effectiveness of controlling documents. Does the time and effort spent controlling documents far outweigh any possible problems should there be a loss of document control?

Clause 4.4.6: Operational control

The Standard requires identification of *operations and activities that are associated with the identified significant environmental aspects* of the organization – *in line with the environmental policy, its objectives and targets.*

In effect, procedures are required to control and verify all functions, activities and processes which have or could have, if uncontrolled, a significant impact (direct or indirect) on the environment.

Environmental impacts of the organization's suppliers come under the controls exercised under this sub-clause, as do those of contractors coming on-site or contractors used by the implementing organization. This is especially so if the suppliers' or contractors' methods of working are known to conflict with the organizations' environmental policy. Contractors may not be aware of the environmental consequences of certain actions (for example, dumping their waste into skips without segregation, or operating noisy drills in the evening when the organization may in fact have come to some arrangement with their neighbours not to do so). Many organizations have induction sessions for contractors as a matter of course (for example, on health and safety issues) so environmental awareness could be included in this existing mechanism.

However, the principle of significance should be considered when deciding what controls to put in place. For example, the contractor who comes on-site to clean office windows should receive less attention than the contractor who is laying new cables or drainage pipes on-site.

Organizations with existing ISO 9000 quality assurance 'process control' procedures should avoid merely relabelling some of these process control procedures as 'operational control'. Unfortunately, some organizations have mistakenly seen operational control as an extension to the quality assurance system and have therefore ended up with very descriptive procedures which are unwieldy to use and maintain. At the early stages of systems implementation, such operational controls should only be in place to control the significant impacts. Other operational controls may need to be written later, as their level of significance increases through the continuous improvement process inherent in the system.

Similarities will be apparent in the structure of such procedures, compared to quality assurance systems (for example the 6 Ws as previously described). However, a different philosophical stance is required and instead of operating, testing and controlling for quality specification and tolerances, some rethinking is needed. For example, plant should be operated in a way that ensures that energy usage and pollution are minimized. This may require a more detailed procedure for start-up – when pollution, in the form of airborne particulates, may be greater until the process has stabilized.

Examples of operational control procedures

Suppliers
A procedure for the purchasing department could include a list of the environmental responsibility criteria required of the organization's suppliers. Meeting such criteria would be a part of purchasing negotiations between the organization and the supplier. The objective would be to stipulate grades of material that have low environmental impacts in production, use and disposal.

Of course, the organization cannot merely use environmental criteria and nothing else. As stated earlier, the purpose of the Standard is not to close a business down. The commercial activities of the organization are paramount and must always come first in any decision-making event. If the business closes down then there is no environmental management system!

The procedure could detail that all existing suppliers be sent a questionnaire about their environmental probity as a first objective to obtain initial information. The procedure would then be written along the lines previously described.

Waste
Procedures could describe which waste goes into which waste skip (one skip for plastics, one skip for paper, one skip for other waste etc.) They could also allocate responsibility for ensuring spot checks as a form of audit; ordering empty skips; removal of full ones; ensuring that they are collected by a licensed carrier and taken to a licensed landfill site; and so on.

The *control* exercised here (as required by the Standard) is the segregation of the waste going into the skip. The *verification* here is the checking to ensure the control is effective. Finally, of course, *corrective actions* would be required if the verification showed that the control was ineffective.

Contractors coming on site

The procedure could detail that every time contractors come on site, they are given an information pack containing, for example:

¤ Site contact names and internal phone numbers (for environmental issues)

¤ What to do in an emergency

¤ A list of 'don'ts' – such as not pouring waste chemicals, washings, oils or paints into the site drainage system and not leaving the engine of their vehicle running unnecessarily

They must sign to agree to this and an individual should be named to be responsible for ensuring every contractor is subject to the procedure.

Bunding inspection

Although bunding is built around storage vessels with the purpose of preventing leakage of chemicals and, therefore, avoiding an environmental incident, in reality it may create a false sense of security by obscuring the development of unsafe conditions and inhibiting maintenance. Frequent routine inspections should therefore be programmed. An operational control procedure should describe the regime of how often bunding is inspected and what action to take in the event of damage to a bund wall.

Part of operational control should include emergency response activities (as detailed below).

Clause 4.4.7: Emergency preparedness and response

The Standard requires that three components be addressed by the organization:

1 *Establish and maintain procedures to identify the potential for, and the response to, accidents and emergency situations*, in order to prevent and mitigate the environmental impacts that may be associated with them.

2 Review and revise the importance of learning from incidents. Obviously corrective actions will be taken and results of audits will be

considered after the occurrence of accidents or emergencies (or even 'near misses').

3   Testing of emergency plans should be planned and the Standard indicates that periodic testing of such procedures should be carried out where practicable.

These three components are described below.

Establish and maintain procedures

An organization is well advised to draw up an 'Emergency Plan' or 'Crisis Plan' and to consider different levels of disaster – including the worst-case scenario. The emphasis of such a plan will be placed upon ensuring that the business survives – that there will be the minimum of disruption of service, or supplies of product, to customers. Safety of individuals, staff and others will naturally be paramount but of course the environmental impact must be considered to address the Standard. The two factors (safety and environmental impact) can of course well be interrelated – large volumes of toxic gases will cause environmental damage whilst posing a safety hazard at the same time.

A suitable plan could cover:

- Identifying critical assets – usually major items of plant

- Defining the disaster team and describing responsibilities

- Emergency services – how to contact them

- Enforcement agencies – how to contact them

- Listing communications – including out-of-hours phone numbers

- Holding regular exercises – to test the system

- Setting up a media response team – who is responsible for contacting the media

Examining the function of the media response team further is worthwhile. It is tempting in the event of an environmental incident to offer no comment to the press. However, other parties, who may be hostile to the organization, will comment. This could be disgruntled neighbours, competitors and others who will take the opportunity to repeat rumours or provide media with exaggerated accounts of the incident.

The media will find out, despite efforts to deter them. An attempt to hide issues will usually be uncovered later and will ultimately reduce the credibility of future communications. If, for example, there is a large spillage to a local stream or river with a high amenity value and the organization does not control the media input, the information any interested party receives may be inaccurate at best and hostile at worst. This could influence future relationships with the media. The news-gathering media work to strict deadlines and information is required from whatever source. It is best, therefore, that this source is the organization itself.

Possible emergencies will most certainly have been identified during the pre-paratory environmental review and suitable responses formulated. At the simplest level of response, this may include a list of competent personnel who can be contacted (with alternatives) in the event of an emergency situation. Provision should be also be made for off-site availability of the information needed to contain the disaster in case access to the site is denied on the grounds of safety. The main objectives of the fire service are to control the fire and/or chemical spillage and save lives. They may cordon off the site and prevent access to staff who require telephone numbers (to make contact with other personnel), access to records or emergency procedures, etc. It is worthwhile noting that even if a fire situation is handled correctly by the fire service, environmental problems can be caused by contaminated water used by the fire services. Such water, contaminated by combustion products, may enter canals, streams and drains and emergency plans should take this into consideration.

It is also important to remember that just because risks are low, it does not mean that emergency plans are unnecessary. Without an emergency plan, minor incidents can escalate into major ones.

Review and revise the importance of learning from incidents
In the case of near misses it is important that such potential incidents are recorded and reviewed and not hidden away or merely forgotten. Such inci-dents indicate areas of risk which on other occasions may turn into environ-mental accidents.

Testing of emergency plans

There may be situations where full-scale testing is not practical and thus consideration should be given to desk-top exercises that can be played out. Examples of such testing include:

- Can key individuals be contacted in an out-of-hours situation?

- If staff are injured can relatives be contacted?

- Are keys to certain areas (for example, solvent stores) available out of hours?

- Are vacations or absence due to sickness covered by alternative personnel?

- Can emergency services access the site – day or night?

- If there is ice on the road during winter would this prevent heavy vehicle access – fire-fighting equipment for example?

- Do the organization's fire hydrants function correctly? Are they maintained and tested?

- Does the fitment on the hydrant fit the fire services hoses?

- Toxic gases may be released during a fire. If so, what are these gases? In which direction is the prevailing wind? What is the likely area that will need to be evacuated for safety as well as environmental reasons? What information will be given to the local police in such an event?

The time to put your emergency plan to the acid test and seek the answers to the above is not the day you have to consider such scenarios for real!

From desk-top simulations to fire drills or full-scale exercises, it would be prudent for any organization to assess the need for such tests so that it can justify its approach. One option is to use the concept of risk assessment (for example, 'likelihood versus consequences' as discussed earlier).

Clause 4.5: Checking and corrective action

As with any human undertaking, errors will occur and systems need to be in place to check that events are happening as they should and that, if errors do occur, suitable corrective actions are taken. The following sub-clauses of the Standard require such activities to be carried out in a structured fashion to ensure an effective environmental management system.

Clause 4.5.1: Monitoring and measurement

This requirement of the Standard is for an organization to *establish and maintain documented procedures to monitor and measure (on a regular basis) the key characteristics of its operations and activities that can have a significant impact on the environment*. This could include:

- Procedures for weighing wastes to landfill

- Inclusion of environmental equipment (such as gas and water analyser) into the inspection and testing procedure (for example, the calibration procedure of an existing ISO 9000 system)

This, of course, links in with any management programmes that have already been established to measure targets which enable progress of objectives to be tracked.

There is also a requirement to review periodically compliance with relevant environmental legislation and regulations. This is a requirement which links to clause 4.3.2 *Legal and other requirements* to ensure that breaches of such applicable legislation do not occur.

Clause 4.5.2: Nonconformance and corrective and preventive action

The Standard requires that *the organization shall establish and maintain procedures for defining responsibility and authority, for handling and investigating nonconformance, for taking action to mitigate any impacts caused and for initiating and completing corrective and preventive action*.

Thus the organization must have the capability of detecting nonconformances and then setting up mechanisms for correcting each nonconformance. Further, it should be able to put into place systems that will prevent a recurrence of the same nonconformance.

A moment should be taken here to consider what is understood by 'nonconformance' because not every organization knows what an environmental nonconformance is. Some time should be spent defining the term 'environmental nonconformance' to all interested parties.

For example, not following a procedure is an easy nonconformance to identify. However, this is really a management system nonconformance. A true environmental nonconformance could be a cracked bund wall that has not been repaired as programmed – thereby increasing the risk of an environmental incident.

The requirement of the Standard goes on to state that any corrective or preventive action that is taken shall be appropriate to the magnitude of the problem and commensurate with the environmental impact encountered. This is to ensure that the organization is mindful of that word 'significance'. Referring to the cracked bund wall above, suppose that it is designed to contain only process waters of low toxicity. Clearly, a similar cracked bund wall intended to contain a cocktail of waste solvents has a much higher priority on the corrective action agenda. Higher priority must also be given to monitoring and maintenance costs than the former bund wall. Therefore the organization is using its limited financial resources in the most effective way.

The closing paragraph of this sub-clause merely calls for an organization to record what corrective actions were taken and to make revisions to any relevant documented procedures. Again, using the above example, a possible change to the maintenance procedure would be to increase the frequency of inspection of the higher-risk bunding.

Clause 4.5.3: Records

This sub-clause states *the organization shall establish and maintain procedures for the identification, maintenance and disposition of environmental records.* These records shall include training records and the results of audits and reviews.

Records need to be kept by any organization to demonstrate that previous activities have been in compliance with legislation, including:

* Discharge consents to sewer

* Process authorizations and variation notices

- Controlled waste consignment notes

- Special waste transfer notes

Other necessary records include: accidents, incidents, training, calibration, performance monitoring, internal audits, training, management reviews.

For example, due to packaging waste regulations in the UK, records such as weigh-bridge tickets will have to be kept as the organization needs to demonstrate to the Environment Agency that recycling and recovery of waste are meeting legislative requirements. The transfer note system requires that both parties keep a copy of the transfer notes and the description of the waste for a minimum period. There may be an occasion when the organization has to prove in a court of law where the waste came from and what they did with it. A copy of the transfer note may also need to be made available to the enforcement authority if they wish to see it.

The Standard also requires that records are legible, easily identified, easily retrievable and protected from damage or loss. The minimum length of time that they are kept should be documented. The approach taken by most organizations is to review what records they need to keep (not all records need to be kept but any record that has an environmental implication should be reviewed as to whether it needs to be retained); where they are kept; and for how long. A procedure can consist merely of a list detailing the above; this has the advantage of being easy to maintain and audit both by internal and external auditors.

Records, of course, also refer to electronic storage. A means of protecting such information through regular back-ups and storage off-site *must* be established. As some of the information should be 'readily available', elaborate password systems must take this into account by allowing access to those who are authorized.

Clause 4.5.4: Environmental management system audits

The Standard requires that an organization carry out *periodic environmental management system audits* in order to:

a) Determine whether the environmental management system conforms to planned arrangements (controlling and minimizing the significant environmental impacts) and meets the requirements of the Standard.

b) Provide feedback to management of the results of such audits.

Such auditing should be performed on a planned and scheduled basis to reflect the environmental significance of the activities being audited.

Audit methodology

Considering the first part of the sub-clause there are two components:

- Effectiveness auditing

- Compliance auditing

Effectiveness or *effects-based auditing* is all about checking that the environmental management system is delivering the improvement in performance that it is intended to by auditing whether environmental objectives and targets are being met. This can also be called a *vertical audit* as it follows a line (the audit trail) right through the entire environmental management system.

Compliance auditing is all about ensuring that procedures are being followed in order to comply with the requirements of ISO 14001.

Effectiveness auditing is certainly the focus as far as certification bodies are concerned. Accreditation criteria guidelines, which should be followed, point the certification bodies towards scrutinizing the internal audit for effectiveness auditing.

Effectiveness auditing

Effectiveness auditing is all to do with verifying that the system in place is operating the way it should and that the objectives set for performance improvement are being met. The easiest way of verifying this is to perform an effectiveness audit which cuts right through a cross-section of the environmental management system.

The internal auditor can start with the list of significant environmental impacts and choose one example. This significant impact should have an objective (or objectives) set against it. Each objective must have some measurable target (or targets) – otherwise an organization would not know whether it was achieving anything.

Thus the auditor will begin to follow the audit trail from targets to a management programme, as below:

Step 1)   Choose a significant impact – check the methodology used to determine significance.

Step 2)   Review the objectives set for that impact. This should highlight some measurable targets

Step 3)   If applicable, check that all of the legislative framework associated with the impact has been identified, communicated and understood by those who need to know.

Step 4)   Ascertain whether the appropriate environmental records are being kept. There should be mention of targets in the management programme – how they are measured and monitored.

If it is discovered that target figures are not being reached and, therefore, that it is unlikely that these objectives will be met, see if this has been addressed via corrective actions by those personnel in control of such activities.

Step 5)   Look at any operational controls that are in place to control and minimize the impact.

Step 6)   Determine whether appropriate awareness is apparent and, if not, whether further training is required.

Step 7)   Consider whether this significant environmental impact has linkages with the environmental policy. Remember – the environmental policy is the driver of the system and therefore must be strongly linked to significance of environmental aspects.

Compliance auditing

This a relatively straightforward exercise ensuring that personnel are following procedures: taking measurements when and where they should; reviewing and updating legislation; generating the appropriate records, etc. Such auditing methodology will be second nature to an organization that already has a documented quality assurance system meeting the requirements of ISO 9000.

Some organizations keep the two audits very distinct – performed at different occasions and by a different set of auditors. Others use the quality assurance auditors for compliance auditing and only use their specifically-trained environmental auditors for the effectiveness auditing. Still other organizations use a combined approach with a team of internal auditors or issue their auditors with basic 'environmental checklists' compiled by an environmental specialist. There are no rights or wrongs here – it is whatever suits the culture of the organization best.

This compliance style of auditing can be called the *check-list* or *tick-box* style of auditing. This has its place in management systems but is of less importance to the third-party external auditor – the emphasis will be very much on the effectiveness auditing.

Environmental nonconformance

It is important that the internal auditors can identify 'environmental' noncon-formances during the audit (as distinct from 'compliance' nonconformance which would be identified as in 'compliance' auditing, above). A compliance nonconformance is, for example, when a member of personnel neglects (due to an oversight) to log onto a register the daily amounts of one waste stream going to landfill. The action is taken but not recorded. An environmental noncon-formance could be a measurable target (such as the monthly figure for use of recycled paper not increasing as planned). If this has not been identified by operatives and management, for whatever reason, and corrective action not taken, then this may affect the objectives set. It may jeopardize a statement within the environmental policy and constitute a potential environmental noncompliance.

Auditor qualifications

Some organizations use existing and available quality systems auditors from an existing ISO 9000 system for the compliance auditing (as outlined above). There should be no difficulty with this practice as these auditors will be well able to audit for compliance against documented procedures. Qualifications for effectiveness audits are different. The auditor would need to have a grasp and understanding of the Standard and the environmental management system, and a broad understanding of environmental issues. Such requirements can certainly be achieved through a independent learning process with a combina-tion of formal training and direct experience (perhaps initially under the guidance of consultants).

The Annex to the Standard also points out that auditors should be reasonably independent of the area or activity that is being audited. Again, this is only common sense. An auditor, auditing his own area of competence, is hardly likely to be impartial – especially if faced with a potential nonconformity directly traceable to their error!

Reporting back to management

As noted initially, the Standard calls for some form of feedback to management on the results of the audits. In truth, this is common sense, because if the results of the audits demonstrate major discrepancies in what was planned (through objectives and targets) and what is actually being achieved, then management needs to reconsider the effectiveness of the whole system in order to fulfil its obligations as set out in the environmental policy.

Auditing procedure

The methodology for performing the audits should be established within written procedures. How else can an internal auditor know how to conduct the audit? Frequencies of auditing should be specified – written in a schedule, plan or even a chart – and this should take into consideration the results of previous audits. Many nonconformities raised at the last audit should trigger off a more frequent re-audit until it is established that the corrective and preventive measures have worked.

Finally, other types of audit used by organizations include external site audits – housekeeping (litter etc.), visual impact from the outside, odours, dust. These additional audits can add to the effectiveness of the compliance and effects-based audits.

Clause 4.6: Management review

The organization's top management shall, at intervals that it determines, review the environmental management system to ensure its continuing suitability, adequacy and effectiveness.

Again, common sense dictates that once a system is implemented, there should be a review process to test whether what was planned does happen in reality. There is no correct way to perform an environmental management review – it must suit the organization's culture and resources. As the Standard refers to 'top' management, this does indicate that a certain level of seniority of personnel should be present at such reviews, to demonstrate commitment. However, there are certain minimum areas to be reviewed and one option, used by most organizations, is to have a standard agenda for each meeting. The first point on the agenda should be a review of the Environmental Policy. This is the 'driver' for the whole system. Senior management should be able to examine it and say with confidence that what was planned (say twelve months ago) as stated in the policy, has occurred or that substantial progress has been made. Thus a typical agenda could be :

• Are the objectives stated in the Environmental Policy being met?

• Does the organization have the continuing capacity to identify environmental aspects?

• Does the system allow the organization to give a measure of significance to these aspects?

• Have the operational controls that were put in place achieved the desired levels of control?

• Are effective corrective actions taking place to ensure that where objectives are in danger of slippage, extra resources ensure a return to the planned time-scale?

• Are internal audits effective in identifying nonconformances?

• Is the environmental policy sufficiently robust for the forthcoming twelve months?

It could be argued that during the early months of the implementation period (perhaps prior to certification) these cyclical reviews are not appropriate and they should focus on just the progress of the implementation of the system. This is a reasonable viewpoint but, as the system approaches maturity, a review as above is beneficial at intervals of six to twelve months.

Accreditation criteria guidelines do indicate that some form of management review is essential prior to certification being granted by the certification body. Of course, such reviews should be minuted, documented and filed for future reference.

If it is concluded that the set objectives are being met, the organization is well on its way to minimizing its significant environmental impacts and thus complying with the requirements of the Standard.

Summary

The requirements of the Standard can seem daunting at first, but by obtaining commitment from top management, with methodological planning and a good understanding of the concepts of ISO 14001, the implementation need not be too difficult and should be well within the reach of the smaller organization.

However, not only does this achievement require management commitment during the implementation phase but, arguably, more commitment is required after the ISO 14001 certificate is obtained. The process of certification and what follows after certification is described in the next chapter.

Chapter 4

The assessment process

Introduction

Unfortunately, it is not sufficient to declare to the world (following implementation) that you have an environmental management system that meets the requirements of ISO 14001. It would be somewhat naive of an organization to believe this. In the real, commercial world compliance with ISO 14001 has to be proven.

One accepted method of validating your claim is to have your environmental management system audited by an external body, one which can demonstrate total impartiality and is seen to be independent by the world. Such an independent body would be an accredited certification body. Assessment and certification of an organization's environmental management system provides a balanced, impartial measurement and verification of:

- Compliance with environmental regulations

- Commitment to care of the environment

- Management of environmental risk

The granting of a certificate to successful applicants and the subsequent registration demonstrate to all interested parties that the organization is committed to environmental responsibility.

In practical terms (and until a better method is discovered), an auditor – or more likely, a team of auditors – physically 'measures' your system against the Standard using interviews, discussions and reference to objective evidence, such as documents and records. After weighing up all the evidence, he will make a judgement as to whether your system satisfies all elements or clauses of the Standard.

Such an assessment process for environmental management systems is adapted from the more mature discipline of quality assurance certification (ISO 9000), which has been operating world-wide for many years.

Certification bodies themselves have their conduct audited by 'accreditation bodies' (see Appendix I) who are, in turn, subject to rules, regulations and controls operated by a higher authority (usually an arm of the national government). These controls exist to ensure that the certification body demonstrates impartiality, the correct methodology, and the right level of competence in the conduct of the certification process. Such a certification process can then withstand scrutiny from all interested parties and has international credibility.

The regulatory framework for certification bodies

In the UK, for example, UKAS (United Kingdom Accreditation Service) is the higher regulatory body governing the activities of those certification bodies offering accredited certification to ISO 14001 (see Appendix I for explanation of 'accreditation'). Their experience in such matters has been gained through quality assurance – ISO 9000 (formerly BS 5750) – which was valuable in setting accreditation criteria to enable accredited BS 7750 certification to occur. This was followed by ISO 14001 accreditation.

Harmonization of accreditation criteria – that is, an agreement on accreditation processes – has occurred within Europe. Seventeen countries are represented via the European Accreditation for Certification (EAC) forum. This includes UKAS which has contributed experiences gained from accreditation of BS 7750.

Further work is continuing across the world to bring global harmonization to national accreditation bodies. For example, the UK has UKAS; the United States has RAB (the Registrar Accreditation Board, which operates jointly with ANSI, the American National Standards Institute); Japan has JAB; Australia has JAS-ANZ; and so on. (See Figure 4.1.)

The considerable effort that has gone into achieving global harmonization is a reflection of the concern that ISO 14001 be interpreted as uniformly as possible,

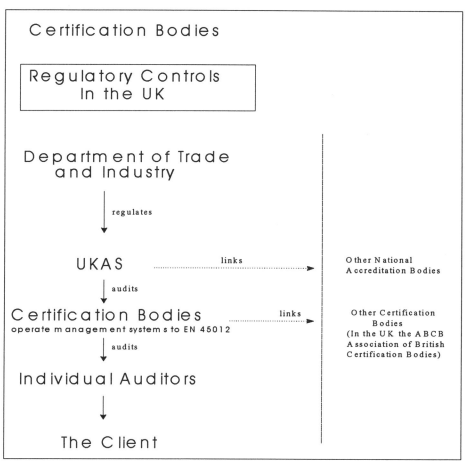

Figure 4.1 Regulatory controls in the UK

thereby ensuring uniformity of auditing from country to country, continent to continent.

Accreditation Criteria (see Appendix IV for summary points) is designed to guide certification bodies towards uniformity of approach. Although designed for certification bodies, an implementing organization would do well to examine the full document as it will help to provide an insight into what the certification bodies will be looking for when establishing the conformity of an environmental management system with the Standard.

The document describes the multi-stage approach required, as well as the certification bodies' management structure, auditor training and competence. The 'characteristics ' and 'qualifications' required of an external auditor, as well as the approach and methodology, are further examined in Chapter 7. The document can be considered to be an 'overlay' specification on top of ISO 14001, which is mandatory for auditors and certification bodies alike, and whose function is to ensure consistency of auditing approach.

The multi-stage approach

The multi-stage approach to certification has not come about by accident. It is based on the considerable experience gained by the certification bodies of quality management systems such as the ISO 9000 series. The process has been designed to ease certification (different from making it easy!) and has been found to operate extremely well for clients. The rationale is that at each step the client has the opportunity to rectify problems and address omissions prior to the full on-site assessment visit – the certification audit. Clearly it would be in no one's interests if a team of auditors arrived on-site to perform the certification audit and, within a short space of time, found so many omissions and errors in the system that the audit had to be aborted. While this demonstrates that certification is not achieved without some effort, it also wastes time, effort and money (the client's) and demoralizes the staff who have probably put much individual effort into implementing the system.

The multi-stage approach is a modification of the quality management system approach, which is:

stage 1)   Desk-top study of the quality assurance management system documentation

stage 2)   On-site assessment

This has been further developed for environmental management systems to become:

stage 1)   Pre-audit (sometimes referred to as the initial assessment)

stage 2)   Desk-top study of the environmental management system documentation

stage 3)   Certification audit (sometimes referred to as the main assessment)

This approach is illustrated in Figure 4.2.

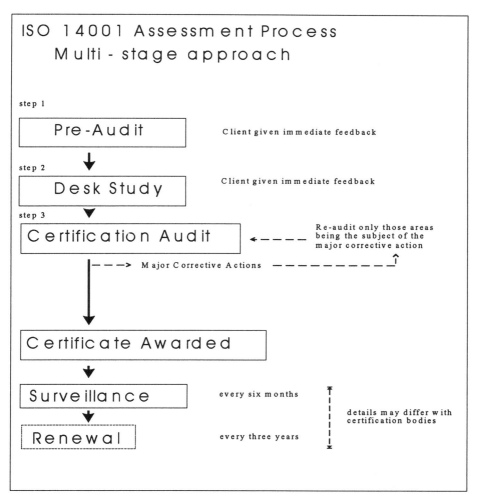

Figure 4.2 The multi-stage approach

This development has come about because of operating experience gained from quality assurance assessments over many years and initial experience of assessments to the 1992 version of BS 7750. Accreditation guidelines for certification bodies state that a minimum 'two stage' approach should be taken unless there is compelling justification for not doing so (see Appendix IV).

It may be that at some point in the future, this requirement for a pre-audit, followed by a certification audit, will be relaxed due to operating experience and the rising awareness of what is expected of implementing organizations. Whatever the outcome, most organizations who have been successfully certified have valued the two-stage assessment approach.

The certification steps are described in detail below.

The pre-audit – objectives

The objective of this stage of the assessment process is five-fold:

1   To ensure that the Environmental Management System is based soundly upon the identification of environmental aspects

2   To ensure that the internal audit methodology is correct

3   To determine the organization's state of preparedness for the next stage of the assessment process

4   To plan and allocate the resources for the next stages of the assessment process

5   To collect necessary information regarding processes of the organization

These objectives are described below.

Ensure that the Environmental Management System is based
soundly upon the identification of environmental aspects
The adequacy of the PER (where applicable) is verified and evidence that the environmental management system takes the findings of the PER into account is sought. Reassurance is sought that a sound and logical methodology is being used to: identify environmental aspects; determine a measure of significance; and apply the resources of the organization to address such significant environmental impacts.

The reason for this examination of the fundamental basis of the system is based upon experience of auditing ISO 9000 systems. It was clear that many organizations worked very hard to create a system to meet the requirements of that standard. However, some were ill-advised and developed cumbersome bureaucratic systems. When it came to assessment, these organizations either failed to comply (because the important points were lost in a morass of detail) or achieved a very successful pass despite the fact that the system was groaning with the effort required to maintain it.

Comments such as 'it requires a full-time employee just to keep up to date with the paperwork' were not unusual and clearly indicated that something had gone amiss. The organization had misunderstood the fundamental purpose of a formal documented management system.

Thus, the purpose of the pre-audit is to discuss with the client the methodology (the foundations) and the decision-making processes behind the documentation. Indeed, during this phase the majority of time may be spent on one long interview in order to test the decision-making processes. Manuals, procedures, work instructions etc. will be briefly reviewed (rather than clause by clause auditing to the Standard) – mainly to see if the client is taking the correct approach. It can be argued that the manuals, procedures and work instructions are not highly important at this stage. Manuals and procedures can be written or amended relatively easily and errors corrected via the word-processing software. What cannot be easily amended is a top heavy, overly bureaucratic system. Furthermore, what is harder to amend or reverse is the damage done to the motivation of other personnel within the client's organization who may not yet be part of the 'critical mass' of personnel driving the cultural changes – they may feel that the system is just a pointless paper-chase.

Ensure that the internal audit methodology is correct
At this early phase, the external auditor needs to examine whether the internal audit is conducted along the effects-based auditing methodology, and whether it is capable of identifying and ensuring correction of environmental nonconformities.

If an organization has an existing quality assurance system, the external auditor will be mindful of the fact that the concept of 'self-policing' may be within the organization's culture, but that the different methodology required by ISO 14001 may not be. The compliance style of auditing quality assurance systems is a 'tick box' approach which does not easily translate into performing effective audits within an environmental management system. These two very different auditing approaches were discussed in previous chapters.

The auditor will need to ascertain whether the most significant impacts are being audited as a priority. The auditor does not expect to see all areas and procedures of the organization audited prior to the pre-audit. However, if it is believed that an area of significance has not been audited, or has not been scheduled according to its order of significance or importance, then this may result in a *development point* or *observation* being raised. The intention of the development point is to direct the organization towards reconsidering how they schedule their audits.

One area in which development points are raised in many organizations, concerns audit schedules showing planned audits covering the next twelve-month period. Such schedules may assign the numerical order of the clauses of the standard to consecutive months: clause 4.1 to be audited in Jaunary; clause 4.2 to be audited in February, etc. Alternatively, audits may be scheduled to occur just before the six-monthly surveillance visits by the certification body (see section on *Surveillance audits* later in this chapter). Both these methods of scheduling audits are flawed and do not demonstrate auditing frequency linked to environmental aspects of significance.

However, an audit schedule 'guide' may be sufficient in the early phases of implementation until experience and knowledge of the organization's environmental impacts is gained. The audit's schedule should be based on order of significance of environmental impacts and any past incidents of nonconformity in certain areas.

Whatever the method used, the company must be able to explain to the third-party auditor the reasons and rationale for the method of scheduling employed.

Determine the organization's state of preparedness
for the next stage of the assessment process
The auditor must gather evidence to show that the environmental management system is designed to achieve compliance with regulatory requirements and policy objectives: in particular, continual improvement of environmental performance. This may entail a brief review of any procedure written to address this requirement.

The auditor also needs to look physically at the site – both internal, and external, to the buildings. The amount of walking about the site and asking detailed questions will be minimal. No depth of audit is performed at this stage. Clearly, the objective of the auditor is to obtain an overview of the site (its size, complexity, local environs and so forth) and to establish that the environmental policy appears to be relevant to these site activities.

Furthermore, the auditor may well ask if the organization has followed any applicable guidance from a *Sector Application Guide*, or indeed, taken heed of the information available in the Annex to the Standard. Questions on whether ISO 14004 has been used as guidance material may also be asked. Following such guidance is not compulsory but if the advice has been ignored the auditor may well ask why.

From the above considerations, the auditor can predict whether the company will stand a fair chance of success under the more intensive scrutiny of the certification audit. If fundamentals are in place, the auditor will recommend that the organization move on to the next stage, providing that on the day of the certification audit there is evidence that the system has been operational for approximately three months. There will be occasions when companies go astray during their initial implementation phase. Such miscalculations should not be discarded but offered in evidence to the auditor. The auditor only requires evidence that the organization has learnt from this and applied corrective actions for the future. The auditor will only be concerned when planned events (that is, targets) are behind scedule and no corrective action has been taken by management. Of even more concern, is when such lack of progress in reaching a target has not been identified. This demonstrates a fundamental lack of control within the management system.

Plan and allocate the resources for the next stages of the assessment process
The next stage of the assessment process is the desk-top study of the system documentation followed by the certification audit.

The auditor will set a date, convenient to both parties, when the detailed desk-top study of the client's documentation will take place. The client may need to make alterations or additions following the outcome of the pre-audit. The amount of work involved here will dictate the date, to some extent.

Planning for the certification audit has several aspects including setting a suitable date. The client in many respects controls this date, as only they know the resources available to address identified areas of weakness before the assessment process can move forward.

The skills required of the certification assessment team will also need to be reviewed. This review would be prompted by the discovery during the pre-audit of more complex environmental aspects than was originally understood by the auditor who performed the desk-top study. This could also arise if the client organization has not fully understood its own environmental aspects and not documented them. These aspects would then, of course, not be 'visible' during the desk-top study. The auditor on the pre-audit may not possess adequate experience in all of these more complex or additional environmental aspects to perform a fair and valid judgement during the certification audit. It would then be appropriate to discuss this with the client and make arrangements to include a specialist, or technical expert with the appropriate expertise, on the certification audit team.

The required number of days to perform the certification audit is also estimated at this stage. This may have been indicated to the client previously for budgetary purposes so this exercise is a confirmation, or otherwise, of the original estimate. The estimate itself is based on several factors, such as: the number of staff on site; the complexity of the processes and their resultant environmental impacts; and the physical size of the site. Geographical considerations must also be taken into account. Travelling long distances will affect the estimate as certification bodies have to pay their auditors the same whether they are auditing or travelling.

The auditor must also consider the arrangements for future surveillance visits (see *Surveillance* later in this chapter).

Collect necessary information regarding the processes of the organization
Further understanding of the client's manufacturing processes will need to be obtained. This will include where applicable 'flowcharts' of the processes and a diagrammatic plan of the organization's site. This map should show the prosition of any rivers, streams, drainage and sewage systems, effluent and air monitoring positions, etc.

Applicable legal documents, such as process authorisations or discharge of effluent consents, will also need to be briefly reviewed.

The emphasis of the pre-audit is on development of the system and not on finding fault. Its purpose is to ensure that the organization is in no doubt as to its state of readiness for the next stage of the certification process. Any deficiencies in the system are put in terms of development points and written into a report. This report may be given to the client on the final day of the pre-audit.

Guidance is given by the auditor as to how such deficiencies can be remedied thereby allowing a smooth transition to the desk-top study.

The desk-top study – objectives

Some certification bodies perform the documentation review (the desk-top study) at the pre-audit stage. Others treat it as a separate stage between the pre-audit and the certification audit. Both approaches have merits. However, the desk study must be completed before the certification audit.

When the client feels that his documentation adequately describes the environmental management system, and has addressed all elements of the Standard, he supplies this documentation to the auditor. The auditor then checks the client's documentation against all the requirements of the Standard. It goes without saying that unless *all* the clauses are addressed, then the Standard is not being met and progress will be delayed until the deficiency is corrected.

The ease of performing this activity varies. There is no set structure for the documentation and it will vary from organization to organization. However, the auditor must be able to find a corresponding reference somewhere in the client's documentation for each clause and sub-clause of the Standard. Because the Standard is generic, clients may interpret some of the clauses in an unexpected fashion. The auditor will look for the rationale behind this interpretation and, if not convinced, will raise a query on this interpretation in the desk-top study report. A signpost document – the environmental manual – enables the auditor to see how the organization has considered each point in the Standard. More importantly, it gives the organization the confidence that it has indeed covered all the necessary clauses!

A desk study can be performed at the auditor's place of work, or on the client's site. The latter has the advantage of immediate clarification for the auditor of an unfamiliar set of manuals and procedures. Against this the auditor may feel he cannot do a professional job if the facilities offered by the client do not match expectations (for example, adequate office facilities with minimum distractions).

Whatever the venue for the desk study, the outcome is the same – a comparison of the client's documented system against the requirements of ISO 14001.

The auditor may be presented with the most typical structure of documentation. This consists of three broad areas:

1   The management system documentation

2   The legislative framework documents

3   Other background information

These areas are described below.

The management system documentation
This management system documentation comprises:

* The environmental policy

* The environmental manual

* The environmental procedures

* The environmental work instructions (if applicable)

Such documentation should include a list of the environmental aspects of the organization and a list of the significant impacts. Any preparatory environmental review (PER) that is relevant could be included but a PER is not assessed by the auditor. It does not form part of the clauses in the ISO 14001 Standard. Therefore, an organization does not have to provide this to the auditor. However, if the auditor feels that the preparatory review is superficial and does not identify the real environmental issues, and the system is based upon this, then questions will be raised regarding the fundamental basis for the whole of the environmental management system.

The legislative framework documents
The legislative framework documents comprise of (where applicable):

* Any relevant national laws

* Permits and authorizations for processes operated

* Copies of water and effluent permits (abstraction licenses, discharge consents)

* Waste producer registration certificates

* Any relevant correspondence with enforcement authorities

Other background information

Other relevant background information might include:

- Site plans – including topography

- Drainage plans

- Prevailing winds (for nuisance issues or fire emergency)

- Boundaries

- Neighbours (domestic dwellings, farmland, amenity areas)

- Any local areas of protected species – wildlife for example

- Any aquifers

- Any canals, streams, rivers

- Process flow diagrams for major processes

One potential development – and one initial cause for concern – was the spectacle of environmental auditors operating outside their home country and not having knowledge of local environmental laws. However, this is not such a formidable problem as perhaps it might seem at first glance.

During the pre-audit and the desk-top study stages, dialogue with the client will be entered into regarding local environmental laws. Copies of legislative requirements can be obtained and translated, if needs be. For a truly international certification body this is less of a difficulty, as local offices will be able to obtain such legislation readily and make it available to the assessment teams. Within Europe, European Directives tend to make for a very similar legislative framework from country to country. At the conclusion of the desk study the auditor will have:

- A list of queries or clarifications

- A list of omissions

- A check list of the salient points that will be audited on-site during the certification audit

A report is issued to the client, detailing points of clarification or omissions against clauses of the Standard so that they can be readily addressed. Thus, once again, the client has the opportunity to address any deficiencies in the system prior to the certification audit. By reporting areas of omission to the client – or indeed areas in need of clarification – before the main assessment site visit, the chances of success are raised.

At this point, the client dictates the time-scale required to prepare for the certification audit. The client can choose when to move forward (as resources allow). The certification body will be mindful of this and, in practice, organizations have taken from one week to six months to prepare themselves at this stage for the certification audit.

Certification audit – objectives

As described previously, the certification audit will be carried out by a team comprising environmental management system auditors, complemented by environmental experts, whose combined skills will be sufficient for conducting a fair and valid assessment of the client's system.

The objectives of the certification audit are two-fold:

1   To verify that the environmental management system complies with all elements of the standard. A desk-top study will have been performed prior to the certification audit and, if all requirements have been met, the task is to ensure that the organization is complying with its own policies and procedures. This task can be carried out by the auditor interviewing the organization's personnel and checking records and documentation against check lists prepared from the desk-top study.

2   To determine whether the management system is designed to achieve, and is achieving, regulatory compliance and continual improvement of environmental performance.

To achieve these objectives, the auditors will focus on the following areas:

• *Identification and evaluation of environmental aspects*: Is the system capable of identifying new environmental aspects and giving them a measure of significance?

- *Consequent objectives and targets*: Are there strong linkages between significant impacts and objectives? Does each significant impact have objectives set against it?

- *Performance monitoring, measuring and reporting, and review against the objectives and targets*: Are procedures being followed to ensure such regular vigilance?

- *Internal auditing and management review to ensure the system's effectiveness*: Are audits being carried out to a plan or schedule? Is the schedule focusing on the most significant impacts first? Are checking and corrective actions taking place?

- *Management responsibility for the environmental policy*: Does the policy mirror the site's activities?

- *Legislative and regulatory requirements*: Are the relevant operational controls in place and robust enough to prevent breaches of legislation?

The step-by-step approach should ensure that any problems at the certification audit are minor in nature and will not delay certification to ISO 14001.

Conduct and progression of the assessment

Having described the three stages of the assessment process and their objectives (see Figure 4.2), this section describes the management and the formalities of the on-site assessment and how it will proceed. As the assessment process is a multi-stage one, each individual stage has its own protocols and each is described below. Again, there may be minor differences in how each certification body performs but such differences will be few and should not affect the integrity, purpose or outcome of each of the stages.

The pre-audit

All certification bodies will operate to broadly the same protocols when it comes to managing the on-site assessment. Therefore, only a generic outline of the pre-audit will be described, as details may be different.

The team leader, who will be a lead auditor from the certification body, will send an itinerary to the client, drafting out the programme for each day of the pre-audit. This programme is open to some discussion with the client, particularly if it does not suit the client's business hours or availability of key staff.

The itinerary will refer to an opening meeting at the commencement of the pre-audit and a closing meeting at the end of the pre-audit.

The pre-audit opening meeting

At the opening meeting, it is expected that the client's senior management will be present. This is the forum for the team leader to introduce the audit team to senior management (and vice versa) and to explain several points of the conduct of the audit. The itinerary is discussed and if there are conflicts of timing etc. then these can be accommodated by the team leader. It is best to keep to the logical order of the itinerary. However, minor alterations to such things as lunch times, or conflicts with essential business activities, are understandable and can be worked around. The auditing team ensures that the organization's personnel are kept informed as to the audit's progress. This includes informing the client immediately if major problems arise which could preclude the award of ISO 14001 (that is, major corrective action requests).

A typical opening meeting would have on the agenda:

* *Confidentiality*: All auditors have signed confidentiality agreements with their employer (the certification body) stating that they will not divulge information (gained during the assessment) to any third party.

* *Explanation of the purpose of the pre-audit*: The pre-audit is not a pass or fail exercise. It is intended to bring any fundamental problems to the surface so as to pave the way for a smooth transition to the next stage of the assessment.

The two fundamental facts that the auditor needs to ascertain are:

i)   That the system is based upon the identification of environmental aspects

ii) That the system is auditable

The auditor will make a note of the personnel who attend the opening meeting and their respective positions in the management structure.

Pre-audit progression

The pre-audit will then occur along the lines detailed in the previous section, ensuring that all five of the objectives are addressed (see *The pre-audit objectives* above).

The pre-audit may in fact only be carried out by one auditor. This depends upon the complexity of the client's operations. Bearing in mind that the pre-audit is a fact-finding exercise and, mindful of the commercial aspects of certification, one auditor may be sufficient. The auditor's knowledge of the client's business and processes may be very superficial at this stage. Any knowledge is based upon phone calls, completion of questionnaires, perhaps a brief meeting, an examination of the environmental policy and a background knowledge of the sector of business within which the client operates. In this way, more detailed information will be gathered by the auditor as the pre-audit progresses.

The pre-audit closing meeting

At the conclusion of the pre-audit, the auditor or auditors will pool their findings and reach agreement as to whether to:

a) Recommend that the company should proceed to the next stage (that is, the desk-top study)

or

b) Recommend any extra work needed before the next stage.

The client is given a written report of the recommendations and development points at this meeting. The client therefore can address any shortcomings as quickly as possible.

The certification audit

The *certification audit* has some processes that are very similar to those of the pre-audit (for example, opening and closing meetings) and these will not be repeated in detail. However, there are some differences, and these will be focused upon. The certification audit is a more structured audit, and compliance is being sought. The auditor cannot offer advice or development points – as was the case during the pre-audit.

Prior to the certification audit, the team leader will send to the client an itinerary, or programme, showing the intended auditing activities. However, the programme will be structured in a way that allows the auditor to audit the requirements of the Standard by following 'linkages' in the system. Only minor amendments will be allowed (to fit in with the organization's start and finish times, for example). The itinerary will again refer to an opening meeting at the start of the audit and a closing meeting at the end of the audit.

The certification audit opening meeting

As at the pre-audit, the opening meeting will be a forum for the team leader to explain how the certification audit is conducted and how it differs from the pre-audit. Additional team members, such as a technical expert (referred to in the pre-audit objectives), will be introduced, if necessary. The itinerary is discussed and, as at the pre-audit, minor alterations can be accommodated by the team leader although it is best to keep to the logical order of the itinerary. The overall objective is to ensure that the client is kept informed of the audit's progress by the audit team. Areas of concern will be discussed with the client throughout the duration of the assessment, so that, near the end of the certification audit, the client should have a fair idea of what the final verdict will be. There should be no surprises at the closing meeting.

A typical opening meeting would have on the agenda:

- Discussions of development points from the pre-audit.

- Discussions of findings from the desk study.

- Informing the client that if problem areas are noted by the auditor, they will be briefly discussed at the time of the finding.

- Discussions of corrective action requests – what they mean and when they would be raised. The significance of and the differences between major or minor corrective actions are explained.

- An explanation of the making of observations – an observation being an item that may assist the organization to improve its environmental management system.

- The explanation that an external audit can only cover a sample of the environmental management system. Such a sample chosen by the audit team will be representative; clearly it serves no purpose to audit 100 consignment notes for correctness if a random sample of ten demonstrates compliance.

- The explanation that when a member of staff is interviewed and does not answer to the auditor's satisfaction, this is perfectly acceptable if the answer can be found in a written procedure or the member of staff directs the auditor to another member of staff who does know the correct answer. Learning the environmental policy by rote is not expected but an understanding of it is.

Certification audit progression

During the desk-top study of the client's documentation, the auditor will have some unanswered queries that can only be answered during the certification audit. These, and other points that need clarification, form the basis of a check list – an *aide-memoire* – for the auditing team.

Such check lists are used by auditors for making notes, not only of areas of potential noncompliance but also of areas of improvement noted since the pre-audit. These notes are also used to provide evidence of the audit. Such auditor's notes are necessary for scrutiny by accreditation bodies to verify the integrity of the audit process. Copious notes made by an auditor do not necessarily indicate problems in the system being audited.

Corrective action requests

Although the purpose of the pre-audit is not to request corrective actions but to raise development points, corrective action requests (CARs) can be raised during the certification audit. They are a fundamental part of the certification process as they ensure continual improvement occurs over a period of time. Corrective action requests should be viewed as a positive outcome of the assessment rather than a negative mechanism for highlighting errors.

That said, the organization must demonstrate that it is following its own environmental policy and procedures. If areas of weakness are discovered then this will be noted. Such areas of weakness will result in a corrective action request being generated by the auditor. There are two categories of requests, depending upon the severity of the noncompliance.

Minor corrective actions

A minor corrective action is requested when a single observed lapse has been identified in a procedure which is a required part of the organization's environmental management system.

The auditor will note the observed discrepancy on a dedicated form. This is handed to the organization's management representative. The management representative signs the document to acknowledge the observed lapse and this document is used to track progress of the corrective action. A minor corrective action would typically be requested for an isolated incident. For example, personnel might note air monitoring data in a register on all appropriate occasions except one. This is plainly a case where a sound procedure is being followed the vast majority of the time but, perhaps for a reason originally unforeseen, it was not followed on this one occasion. Perhaps the member of staff responsible for keeping this register was ill and nobody thought to appoint a deputy to the task. A possible corrective measure to be undertaken by the company would be to name a deputy should a similar situation arise in the future.

Consider another example, this time concerning an environmental improvement target that has been set by the organization. This target is to be reached within twelve months. It could be that after six months, no progress has been made. Internal audits will have identified this lack of progress and although senior management is therefore aware, nothing has been done to correct this potential failure to meet the target. The auditor would raise a minor corrective action request in this instance. The reason for this request is not the failure to meet the target but the failure of the environmental management system to recognize that all was not going to plan and to take corrective actions.

Minor corrective action requests raised during an assessment do not necessarily prevent the organization being certified to ISO 14001. Once a minor CAR has been made, the organization is required to respond in writing (on the form mentioned above) within a specified period of time – typically six months. The organization should detail the actions taken or proposed, in order to prevent recurrence of the problem. The auditor can then respond to the organization to confirm, or otherwise, that the action or proposed action, is appropriate.

This is a mechanism to ensure that the client will take the appropriate corrective action, in advance of the next scheduled surveillance visit – for that is the occasion when the effectiveness of the corrective action can be fully verified by the auditor. It follows that sufficient evidence of the corrective action must have been generated to show the auditor on this visit. If, for whatever reason, there is insufficient evidence to demonstrate the effectiveness of the corrective action, the auditor will be placed in the position of escalating what was a minor corrective action request into a major corrective action request. This has implications for the continuation of the ISO 14001 registration.

Major corrective actions

Major corrective actions requested during a certification audit will preclude the environmental management system from being certified. They act as a 'hold' point. A major corrective action is requested where there is an absence, or a total breakdown, of a procedure that is a required part of the organization's management system.

It could be that at the pre-audit stage, and at the desk-top study stage, although the procedures as written appear to meet the requirements of the Standard, they may not in fact be fully implemented. For example, a small department in the organization may not have been given enough training in operating to the procedures. It may be that there is some personnel resistance to changing the way of working to one of the documented procedures. Whatever the reason, the system is not working as it was planned and documented.

Where an observed noncompliance with documented procedures is, in the opinion of the auditors, likely to cause an environmental incident or accident, then the requested corrective action will always be categorized as major.

Similarly, as for minor corrective actions, the auditor will note the discrepancy on the corrective action request form. However, on this occasion, a much shorter time-scale is allowed for a written response to be verified by the auditor on a dedicated extra visit. Such a visit is an additional expense to the client. Typically, a written response is required within one month and the additional visit within two months from the date of the certification audit. The purpose of this extra visit is purely to verify that the corrective action has taken place and that evidence to demonstrate this is available. If the major corrective action is satisfactorily resolved, then certification can proceed.

The reason for this shorter time scale is to ensure that the organization's momentum towards gaining certification does not falter. However, there should be sufficient time for the organization to strengthen its resources to correct the discrepancy.

It should be noted that major corrective actions requested during surveillance visits are taken very seriously by certification bodies and this is reflected in the even shorter time-scales allowed for corrective actions. The auditor will require a written response within two weeks of the surveillance visit, and will schedule in an extra visit within one month of the surveillance visit regardless of whether written notification of action has been received or not.

It could be that, during the certification audit, the failure of the organization to meet an objective for a significant environmental impact on one occasion may be worthy of a minor corrective action request. For example, on this one occasion a manager made an error in figures or overestimated available resources. However, if the organization persistently fails to meet its own objectives and targets, this rather suggests that its methodology and rationale for setting such objectives is flawed. This would indicate to the auditor that a fundamental part of the system was not capable of delivering system improvements. In this case, a major corrective action request would be needed to ensure that this part of the system was made more robust. If the management controls are not robust enough to detect and correct such system failures, then how can improvement, which is a fundamental requirement of ISO 14001, take place?

There is much debate about what constitutes a minor or a major corrective action and the author does not wish to be drawn into this discussion. There are guidelines of course (as outlined above) but in many respects the actual categorization depends upon the professional integrity of the team leader, or auditor, on the day of the assessment (or surveillance) who balances all the evidence available and then makes an objective judgement.

Observations
The auditors may also make observations. These are intended to indicate to the client areas of potential improvement and are generally followed through at the next surveillance visit. However, if the organization has considered such advice but decided against further actions – due to reasons that the auditor was not aware of – the auditor will not debate the point. Corrective action requests must be actioned – observations need not be. The common theme running throughout the multi-stage process is that it serves no useful purpose for an organization to fail the certification audit badly. It is demoralizing for the organization involved and demotivating for the auditors. Auditors, like other professionals, thrive on success and successes are not measured by the number of nonconformances raised.

The certification audit closing meeting

At the conclusion of the certification audit, the auditing team will consider their individual findings and reach an agreement. The team leader will decide to do one of the following:

- Allow the company certification unreservedly.

- Recommend certification with minor corrective actions.

- Delay certification until further work on the system is undertaken due to major corrective action requests.

The verdict is given to the client verbally by the team leader who discusses with the client any corrective actions and observations. A fully documented report is issued at a later date describing in detail the progress of the assessment (see later in this chapter).

Potential areas of difficulty during the certification audit

Shared sites

There are occasions when the audited site is shared by another organization and this can give rise to potential difficulties. Such difficulties arise when the organization is not able to control the environmental activities of another organization that rents or leases part of its site. This might prompt a 'stakeholder' to question the value of ISO 14001 certification. For example, consider the leased organization causing a pollution incident by an out-of-control discharge of effluent into a river. If this same river runs through the certified organization's site, it could be argued that they, the certified organization, have some responsibility to ensure that such discharges do not occur.

The auditing team will consider the guidance offered by *Accreditation Criteria* (see Annex IV). This states that all interfaces with environmentally relevant services or activities that are not within the scope of the environmental management system – and how they are controlled and influenced – should be appropriately addressed within the scope of the system. Such interfaces, include the 'shared' river referred to in the previous example or perhaps shared sewage and drainage systems.

In practice, the assessed organization must demonstrate that lines of communication do exist between the two organizations. It must also demonstrate that some measure of significance has been allocated to the environmental impacts of the other organization on the shared site. Such interfaces include allowing environmental audits to take place of the leased site, for example, and dialogue between both parties taking place if significant impacts are discovered.

Multiple sites

As in all auditing activities, what is audited is, in practice, a representative sample. Auditing environmental management systems is no different. It has to be this way to give a cost effective result. If the auditor is satisfied with the content of a random sample of, perhaps, ten weekly emissions monitoring results out of an available 52, then it is assumed that the other 42 are also satisfactory. If two out of the ten sampled showed some discrepancies, then the sample size would be increased. This is all part of auditor training and methodology.

The same methodology can be applied when auditing an organization that has a number of sites. If the audit goes well at the selected sample of sites, then there is no need to audit the others at the certification audit. This does assume however, that the environmental aspects of all the sites are similar. Sites with totally different significant environmental impacts would all have to be visited, as well as sampling those with similar impacts.

Again, the guidance for certification bodies is that at least one third of the sites should be sampled. The remaining sites must be audited within 12-24 months of the certification audit at the normal six-monthly surveillance visits.

Minimum level of implementation required

At the pre-audit, it may be that the system is so new that the ink may not quite be dry on the documented procedures! This is not an issue provided that the auditor can be sure that the organization will be moving forward with the correct methodology, as previously discussed. However, at the certification audit, the auditor needs to see evidence that the system is implemented and maintained. The minimum level of such implementation could be open either to individual auditor interpretation or to certification body interpretation. Guidance is once again offered by *Accreditation Criteria*:

1   The system must have been operational for a minimum of three months.

2   The internal audit system must be fully operational and be seen to be effective

3   One management review has been conducted

Requirement 1 merely means that records, reviews, measurements, monitoring, training and other activities should have been operational from three months prior to the certification audit.

Requirement 2 means that internal audits must have been programmed and started, and must embrace the correct methodology. If nonconformities have been raised during such audits, the client must indicate what corrective actions have occurred or are intended. Effectiveness can also be demonstrated by keeping records of audits that show the depth, breadth and objective evidence used to support the audit.

Requirement 3 is essentially a demonstration of management commitment to ISO 14001 implementation. One or two progress reviews of the implementation process would be sufficient for this demonstration.

The audit report and the issue of the ISO 14001 certificate

Although a verdict is given to the client by the lead auditor or team leader (as described previously) this is, strictly speaking, a recommendation only. The final decision to grant the award of the certificate must be given by the governing board of the certification body, which will follow protocols established by the national accreditation body. As has been stated before, such national accreditation bodies are working closely together to give a high level of consistency of assessments throughout the world.

The progress to certification from the verbal recommendation depends to some extent on the particular procedures that the certification body employs and the *Accreditation Criteria* guidelines. Essentially though, the team leader submits a report to the governing board giving a clear indication of the implementation of the various elements of the Standard and the effectiveness of the environmental management system with respect to the achievement of objectives and targets. The report also provides details of:

• The amount of time spent on the assessment (the man days), including documentation review, preliminary investigations, pre-audit, certification audit and reporting

- Names and functions of audit team members as appropriate: team leader, lead auditor, technical expert

- Names and functions of those interviewed at the audited organization's site

- Assessed activity and regulatory requirements at the audited organization's site; a brief description of the legislation/regulations applicable

- Topics for which files have been investigated; some objective evidence of audit investigations

- Evidence of nonconformities to the Standard – in relation to both documentation and implementation

- Observations

On review of the report, the certification body's authorized individual (usually one of the directors) will sign the report. This signature is the evidence of the review and allows the ISO 14001 certificate to be generated by the appropriate administrative function within the certification body. However, during the review, there may be some queries. The queries may concern the technical or typographical accuracy or, more importantly, the substance. Typical queries include:

- Was the correct audit team chosen?

- Was the team able to demonstrate that they had the correct breadth and depth of knowledge to enable an accredited certificate to be issued?

- Was the certification body operating within its area of accreditation?

These and other questions will form part of the review process.

Once the certificate is generated, it must, in turn, be reviewed for correctness and signed by the same authorized individual as above. The client also receives a copy of the signed report and then, of course, a copy of the certificate – the goal of the whole process!

Surveillance audits following the assessment

The assessment hurdle is not the end of the process – the client must prepare for ongoing surveillances.

Again, the approach described may differ slightly from one certification body to the next but some form of continual surveillance activity is required by the accreditation bodies. This surveillance activity must be, at a minimum, one surveillance visit every twelve months. However, this must be part of a programme of such visits which include a full certification re-audit every three years. An alternative strategy is to conduct surveillance visits every six months. These visits are organized by the certification body to ensure that over a period of three years the system is audited against every relevant part of the Standard. At the end of such a three-year period, the certification body reviews the overall continuing effectiveness of the organization's environmental management system. The review will cover the number and seriousness of corrective action requests, as well as any major changes to the organization's operations – including those which may require additional time on site at the next surveillance visit. This alternative strategy is the preferred course of action and is often referred to as *renewal* of the certificate.

Whenever possible, the auditor for the surveillance visits will be a member of the team that performed the certification audit. The auditor will therefore be familiar with the organization in question and will use the surveillance time more effectively. Surveillance visits are planned in advance for a mutually agreeable date.

The objective of each surveillance visit is to sample only a part of the total system. However, at each visit the following is always examined:

- The effectiveness of the system with regard to achieving the objectives of the organization's environmental policy

- The level of management commitment

- The functioning of procedures for notifying authorities of any breaches of authorizations or discharge consents

- Progress of planned activities aimed at continuous improvement of environmental performance – where applicable

- The internal audit – methodology, scope, depth and follow-ups of any internally identified corrective actions

- Follow-up of any corrective actions raised by the certification body at the previous surveillance visit (or following the certification audit)

Should there be no areas for concern at the surveillance visit, the auditor will complete a visit report, either hand-written there and then or sent to the client after suitable word-processing. If applicable, the auditor will make recommendations to the client for improvements to the system in the form of observations. Reports from surveillance visits should include as a minimum:

- Clearing of any nonconformities raised at previous visit

- A record of the amount of time spent on the surveillance

- The names and functions of the audit team members

- The names and functions of those interviewed at the audited organization's site

- The assessed activity and regulatory requirements at the audited organization's site

- A description of all nonconformities revealed

- Any observations noted

If the auditor discovers any noncompliances with the client's documented procedures a corrective action request will be raised. This will be followed up at the next surveillance visit, as previously discussed.

It is recommended that, wherever possible, environmental management system surveillance visits are combined with other certified management systems visits (such as ISO 9000 quality assurance). This recommendation comes from guidance documents for certification bodies, with the proviso that the aspects relevant to the environmental management system are clearly indicated. Such 'integration' or combining of visits is discussed further in Chapter 5.

Some organizations may still have BS 7750 certification as a transition period was allowed between BS 7750 withdrawal and ISO 14001 replacement. Surveillance visits can be structured so as to enable the auditor not only to perform the listed surveillance activities but also to make the transition to ISO 14001 certification. For UKAS accredited certificates this will be conducted by a consideration of the 18-point check list produced by UKAS which they expect organizations to address (details in Appendix II). Most of the 18 points are minor and should result in minimal work by the organization and probably no additional expenditure. A 'seamless' transition is the aim for the client. A new ISO 14001 certificate to replace the BS 7750 certificate will be given by the certification body.

Certificate renewal

Depending upon the options chosen or offered by the certification body, some form of 'extra' scrutiny of the environmental management system is required every three years.

This can range from a prolonged surveillance visit to almost a full certification audit visit. As described under *Surveillance methodology*, only a representative sample of the system is audited every 6 months by the auditor. It could be that, after three years, so many changes have occurred that the system, although appearing to be sound on the 'snapshot' type of review of a surveillance visit, may no longer meet all the requirements of the Standard.

Likewise, over a period of three years, many minor corrective action requests may have been raised indicating perhaps a lowering of the level of commitment by senior management to the principles of the Standard. In this case, a virtual full re-audit of the system is required to give the certification body a higher level of confidence. This option is usually included in the wording of the contract agreed between the certification body and the client.

Certificate cancellation

An ISO 14001 certificate can be cancelled for several reasons.

If an auditor finds evidence of deterioration of the environmental management system during a routine surveillance visit he will make either a minor or a major corrective action request (depending upon the severity of the system breakdown).

Minor corrective action requests can be a fact of life. They will apply typically to minor oversights usually related to human error. Major corrective actions requested after certification are taken very seriously by certification bodies as they indicate a fundamental breakdown in the system – an environmental management system which the certification body has previously assessed and passed. The credibility of the certification body itself can be at risk.

If such a situation arises and the major corrective action request is not implemented within the required time-scale (usually one month, although there can be extensions to this) then the certification body will take the necessary steps to recall the certificate and cancel the registration to ISO 14001.

A certain amount of publicity by the certification body could follow so that other organizations and interested parties are informed that the offending organization no longer has ISO 14001.

The procedure outlined above may seem harsh, but the certification body must be seen to be protecting the integrity of the certification process and the value of the certificate. The message must always be that ISO 14001 certification is not easily obtained and, once obtained, an organization must demonstrate sound environmental practice and continuous improvement in order to retain the certificate.

In reality, very few certificates have been withdrawn for the above reasons. When they have been withdrawn, it has usually been due to massive company restructuring. In such cases the survival of a business has had to overshadow all other considerations. In the majority of cases, a major corrective action request from the auditor is sufficient to galvanize the organization into extra activity to bring the system back into line with its policies and procedures.

Other reasons for certificate cancellations are to do with company closure or companies changing their name.

Summary

The assessment process is a highly-regulated, incremental process comprising pre-audit, desk study and certification audit. It is designed to deliver consistency of approach to the client and allow any deficiencies to be corrected along the way. The correction of deficiencies allows the certification audit to have a reasonable chance of proceeding smoothly. Major areas of concern which could otherwise halt certification are thereby avoided.

By explaining the mechanics of the process, it is hoped that organizations about to implement ISO 14001 will know much more about how the certification bodies operate and will feel less daunted by the actual assessment process. The intention of any certification body must be to make the mechanics of the certification process easy whilst at the same time ensuring that where errors or deficiencies occur they are appropriately addressed through corrective action requests. Ongoing surveillance visits ensure that not only must certification be earned, it must be maintained.

Chapter 5

Integration of environmental management systems with other management systems

---

Introduction

The reader may belong to an organization that already has the ISO 9000 Quality Assurance Management System in place. There is a reasonably high chance of this because there are approximately 100,000 ISO 9000 certificates world wide. A sizeable part of this chapter discusses the relationship between ISO 14001 and ISO 9000. This chapter will offer limited guidance in management systems integration – an area undergoing rapid international development. Each organization must make its own choices as to whether to run a business with one or more separate systems or to join them together: integration. Fundamentally, there is no reason why many separate management systems cannot be joined, combined, amalgamated or integrated. However, with a book of this size, the emphasis is on ISO 14001 implementation. This chapter is restricted to only a few examples of integration of mainly British Standards already in existence or being developed. Other countries will have similar national Standards and the concept of integration of such national Standards remains the same. Likewise, this chapter will not list the clauses or requirements of each standard and compare and contrast them with ISO 14001. Instead, this chapter will focus on not only the common features of both systems but also their salient differences. It is important to keep these differences in mind when attempting to integrate one or more standards.

Much development work is being done world-wide to develop such integrated management systems and the ideal goal – which may never be realized – is to produce one definitive 'Standard' for any business. This Standard would address all an organization's activities and would be used as a complete business model for any new organization.

The small to medium-sized enterprise may well hesitate before implementing another management system that will stretch management resources and time further. Such organizations desire a painless grafting, or amalgamation, of their separate systems.

Organizations without ISO 9000 may debate whether they should implement ISO 14001 at the same time as ISO 9000 or achieve certification for one before the other. A typical question is: Does having ISO 9000 certification assist in the process of implementation of ISO 14001?

Other debates centre around:

- The economics of certification

- The best use of auditors' time – both internal and external third party

- Whether it is just as easy to audit two systems internally as one

- Whether costs are reduced because elements of the separate systems overlap

This chapter will address these issues and attempt to answer such questions.

What is an integrated management system?

For the purpose of this book, an integrated management system is a management system comprising of ISO 14001 plus at least one other management system. Both systems should run concurrently with each other in an organization and both should be capable of being audited by an external body to a recognized national or international standard.

An integrated management system is sometimes mistakenly referred to as a 'total quality management system'. *Total Quality Management (TQM)* is very much a philosophy-based model for running a business. TQM encompasses all facets of a business, including changing of the culture of the organization and persuading personnel to think and work in different and better ways. Although there is a British Standard (BS 7850) for TQM, some critics argue that true TQM can never be reached as it is an endless road of continual improvement. Integration of management systems is somewhat removed from this and only considers the following aspects of running a business:

- Management responsibilities and accountabilities

- Business processes

- Deployment of resources, skills, knowledge and technology

These aspects are integrated to ensure that the business delivers its objectives. The objectives of the business include elements of quality, the environment and health and safety. These objectives are the same as many of the requirements of stakeholders (as discussed in Chapter 1):

Customers	*Shareholders*
Safe and reliable products/services	Return on investment
Reliability of supply	Profitable business
Fitness for purpose	Legal compliance
Environmentally safe product	Good image
Value for money	Growth
Employees	*Community*
Safe working environment	Minimum environmental impact
Job satisfaction and security	Employment opportunities
Care and recognition	Stability
Rewards for good work	

In many ways such requirements are not met by just environmental responsibility but a much wider range of employer care and responsibility.

Reasons for integration of management systems

The burden on management time within the organization can be reduced if one element of a management system can be addressed at the same time as the same element of the other system. For example, the environmental management review for ISO 14001 could take place at the same time, with the same personnel, as the management review for ISO 9000.

Fees incurred for the Certification Body to carry out its routine surveillance visits can be minimized by only having the external auditor arriving every six months (that is, twice per year to perform a combined ISO 9000 and ISO 14001 audit rather than four visits, two for ISO 9000 and two for ISO 14001). As suggested in the introduction, a single standard that integrates all elements of a modern management system into one auditable standard is the ideal. However, for the near future, it must be appreciated that individual standards have been developed for different purposes and therefore organizations wishing to integrate their management systems will need to develop their own model for an integrated system.

A question which is often asked by organizations considering ISO 14001 implementation is: 'How well will the new system fit into, or combine with our existing management systems – such as quality assurance, occupational health and safety, data protection and so forth?'

This is a fair question. The organization concerned will have used its limited management resources to attain its existing management systems and will be anxious that installing a new system will not bring new problems to threaten existing systems. For example, will a serious nonconformance in the new environmental management system have repercussions for their existing quality assurance certification? Will they have to have a whole new documented system which will be a duplicate, in volume and complexity, of their existing system?

However, these are isolated questions and conceptual queries. Taking a broader, longer term stance, there are always new challenges and demands to be met when managing any business, especially when viewed against:

- Significant competition

- High customer and community expectations

- Returns on capital employed

- Regulatory compliance

- Executive liability risk

If a business is to meet the challenges above successfully, then it needs to call on all its management resources – especially when fulfilling the requirements of Quality, Health and Safety and Environmental Standards. Integration of these management standards would allow common areas of the standards to be managed, thereby making more effective use of management time.

Looking at the implications and consequences of a separate systems approach:

- Actions and decisions are made in isolation and are therefore not optimal.

- Employees are presented with a proliferation of information and even conflicting instructions which may put the company at risk.

- Bureaucracy can flourish – how many systems can the organization cope with?

- Lack of ownership.

- Wrong and costly decisions are made due to non-optimization of resources.

These consequences can be addressed by developing an integrated approach to quality, health and safety, and environmental management in particular as these three standards cover a high proportion of the scope of any business. Integration will expose areas of waste and no-value-added activity, and provide opportunity for rationalization of:

- Documentation

- Auditing and review processes

- Barriers across departments and functions (by their removal)

The majority of organizations implementing ISO 14001 (or who have gained certification) have an existing ISO 9000 quality assurance management system so the integration of these two standards will be discussed first.

Standards for integration

ISO 9000 series of standards

ISO 9000 is a series of standards for the management of quality but only three of these are designed to be assessed by a certification body:

> **ISO 90001:1994 Quality Systems** Model for quality assurance in design, development, production, installation and servicing.

> **ISO 9002:1994 Quality Systems** Model for quality assurance in production, installation and servicing.

> **ISO 9003:1994 Quality Systems** Model for quality assurance in final inspection and test.

The purpose, implementation and certification of quality assurance management systems will not be entered into here. Due to its long history of development (some 30 years or so) there are many excellent books available covering all aspects of quality assurance. However, some description would be useful at this point for comparison purposes.

Briefly, in order to be successful an organization must offer products or services that:

• Meet a well defined need, use or purpose

• Satisfy customer expectations

• Comply with applicable standards and specifications

• Are made available at competitive prices

• Are provided at a cost which will yield a profit

The ISO 9000 series of standards act as a model for an organization to follows in order to meet these requirements.

ISO 9000 has behind it some 30 years development – from its embryonic days as a military standard for ensuring consistency of product (especially munitions) through sub-contractors providing materials for the government (in the UK as BS 5750) to the adoption of this as an international standard, ISO 9000, in 1987.

BS 7750, the precursor of ISO 14001, was designed to be compatible with such quality systems and its name – BS 7750 – was intended to demonstrate its relationship to BS 5750.

The reason for independent certification of an organization's quality system was one of 'confidence'. By purchasing from an organization thus certified, the buyer would have a high level of confidence that goods (or services) purchased would meet specified requirements. The contract would then be fulfilled between customer and supplier and meet the five points above. Specifications were set that allowed tolerances on product dimensions, weight, colour, quantities etc. that the sub-contractor could achieve in the manufacturing process and at the same time fulfil the customers needs (that is, fitness for purpose). The customer was also assured that if things did go wrong during the transactions, the mechanisms within the supplier's management system would ensure corrective actions would be taken and that the same problem would not recur.

In an environmental management system there is no 'actual' customer setting a specification in a similar fashion. The nearest that an environmental management system has to a customer is of course compliance with legislation. In effect, the national government is setting a specification for the organization to work to. The requirements of the process authorization, with its limits of discharge and levels of contaminants or VOCs are a form of tolerance either side of the optimum. It can be argued that there is, in fact, a spectrum of 'unwritten' customers, that is the stakeholders – in the guise of shareholders, investors, the general public – whose requirements must be satisfied.

There are big differences here in the purposes of the two standards: ISO 9000 and ISO 14001. It would be a mistake for any organization to think that it could 'graft' on an environmental management system to a quality assurance system with the bare minimum of effort.

There are also big differences in the philosophy between the two standards (ISO 9000 and ISO 14001). Continuous improvement is a basic requirement of ISO 14001 and this is explicitly stated in the Standard. The organization has to demonstrate year-on-year improvements in its environmental performance.

ISO 9000 has no such statement of ongoing improvements within its clauses. Some people believe that an ISO 9000 system – via its corrective and preventive action processes and nonconformance system – should be capable of ultimately reducing wastage by reducing errors. Therefore, ongoing improvements will be made. In manufacturing, for example, this will result in the use of less raw materials and energy. Reduction of raw material usage and energy consumption usually feature in any manufacturing organization's ISO 14001 register of significant environmental impacts, so there do appear to be some similarities of intent. This is certainly an area for fierce debate between purists of each standard.

The long history of ISO 9000 provoked much discussion about the value of the Standard, and both positive and negative views have emerged. Most organizations seeking ISO 14001 certification have existing ISO 9000 certification and are aware of the criticisms that have been levelled at ISO 9000. These include:

- Lack of reference to the associated requirements of health, safety and environmental protection

- Lack of focus on continuous improvement

They are also aware that although ISO 9000 has been adopted as a management system in over 70 countries, it really is only an entry level 'qualification' into quality assurance. Models for business excellence and TQM now exist – for example, EFQM (European Forum for Quality Management) in Europe and BS 7850 in the UK.

Similarities between ISO 9000 and ISO 14001 are readily apparent especially in the structure of the documentation required (see Figure 5.1). Both ISO 9000 and ISO 14001 demand a certain level of documentation. *The organzation shall have documented procedures* is a phrase that is common to some clauses in both standards.

Some considerable time and effort was expended by other authors on comparing and contrasting ISO 9000 to the then BS 7750. It was relatively straightforward to put the two standards side by side and compare and contrast them, and to draw up a difference and similarity matrix. Annex B of ISO 14001 does include a matrix of detailed comparisons to ISO 90001.

The Structure of an Environmental Management System and a
Quality Management System have commonalities:

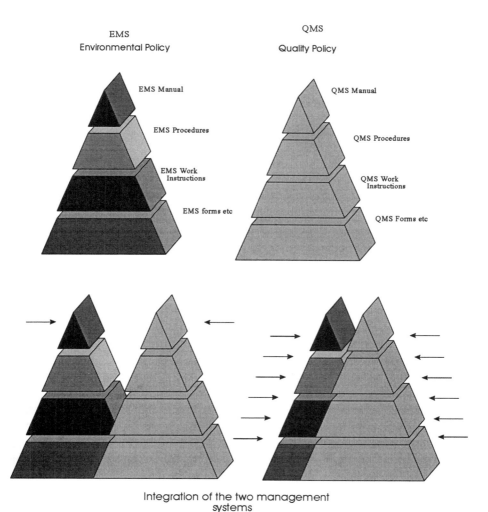

Integration of the two management
systems

Figure 5.1 Management system commonalities

It should always be borne in mind – no matter how many the similarities – ISO 9000 and ISO 14001 are intended for totally different purposes. These different purposes are made apparent when looking at the broad differences as indicated below (the reader can refer to Annex B of the Standard as mentioned above for detailed comparisons):

	ISO 14001	ISO 9000
Philosophy based on identification of environmental aspects and minimization of significant environmental impacts	*yes*	*no*
Commitment to continuous improvement	*yes*	*no*
Compliance with environmental legislation	*yes*	*no*
'Effects' based internal audits of the system	*yes*	*no*
Meeting the demands of a spectrum of stakeholders	*yes*	*no*

There are other 'quality management' systems which have used ISO 9000 as their basis and these can be integrated equally well with ISO 14001.

QS 9000
This standard is of some importance due to the global size of the industry supplying parts and materials to the motor industry. QS 9000 is the result of the world's largest three motor manufacturers – Ford, Chrysler and General Motors – defining their needs from their supplier base in collaboration with each other and requiring that their first-tier suppliers implement a quality assurance system that complies with QS 9000.

QS 9000 itself is very firmly based upon the requirements of ISO 9000 but with additional clauses. In fact these elements originate from the three separate supplier quality assurance schemes operated by the 'big three' and are harmonized into QS 9000.

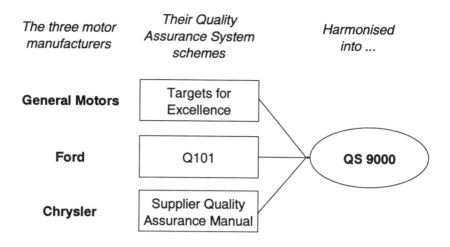

These are now harmonized into one standard with target dates for their suppliers to achieve. If suppliers fail to meet these deadlines they would be removed from the tender lists or approved supplier lists. The main reasons for this co-ordinated approach by the 'Big Three' are based on past experiences of :

- Lack of common terminology within the industry

- Proliferation of similar standards

- Different documentation needs

- Differing audits

- Supplier concerns (that is, it was more costly for suppliers to implement and maintain three quality systems to meet the requirements of the 'Big Three' than one common quality system)

The three companies are looking for reassurance that their suppliers are committed to continual improvement and embrace all elements of good management within their business practice.

Obviously any organization failing to manage its risks in the environmental and occupational health and safety fields may put their production output at risk, compared with those that have effective management systems controls in place. Environment incidents or disasters could stop production. Investigations into accidents at the workplace could prevent machinery from operating. These incidents and accidents would prevent products from reaching the customer at the required, and contracted, time.

An organization with an existing ISO 9000 system will find many commonalities with QS 9000. This is to be expected, as QS 9000 is based very much on ISO 9000. However, QS 9000 has some specific requirements that have a close relationship to ISO 14001. In clause 4.9 of QS 9000, *Process control*, there is a requirement that the organization shall have a mechanism to identify all applicable government safety and environmental regulations, including those concerning handling, recycling, eliminating and disposing of hazardous materials. Clearly this has a direct link to sub-clauses 4.3.2 *Legal and other requirements* and 4.4.6 *Operational controls* of ISO 14001. Therefore, these would be common procedures if the organization was certified to ISO 14001 and QS 9000.

It should also be noted that motor manufacturers are generally looking for their suppliers to implement an environmental management system in the near future.

BS 8800:1996 Guide to Occupational Health and Safety Management Systems
In recent years organizations of all types have been actively addressing the question of health and safety in the workplace, partly in response to legal requirements but also in response to the fear of litigation and the pressures of increasing insurance premiums and other (uninsured) costs. With the ever increasing need to comply with statutory legislation and national regulations, they are realizing that installing a documented occupational health and safety management system can often highlight potential legislation shortfalls.

Traditionally, the approach to health and safety in the workplace has been to have an occasional check on the safety aspects of working conditions – usually prompted by the occurrence of a serious accident. However, audits in themselves do nothing more than give a 'snapshot' of how physical conditions are at a particular moment. They do not give any indication of the inherent safety or otherwise of the operation, or the probability of a serious accident occurring. Nor do they give any significant assistance in the control or management of safety in the workplace. Existing, but perhaps unstructured, systems may be in place.

For example, all organizations in the UK will have been subject to health and safety legislation for many years. They will have had to meet the requirements of the Health and Safety at Work Act (1974). Therefore it is highly probable that there will be members of staff with some experience of ensuring compliance with such regulations. However, it is likely that such systems will be set up with the aim of avoiding safety incidents that might lead to prosecution rather than with the aim of aiding a more effective and profitable business. The driving force will therefore be negative, defensive and reactive.

True assurance on safety can only be achieved in the context of a structured management system, which strives to manage safety in an active way rather than simply to react to events.

ISO 14001 does not require the inclusion of Occupational Health and Safety (OHS) within the environmental management system. It is designed so that Health and Safety may be incorporated if an organization wishes to do so. The legislation covering occupational heath and safety is extensive in most developed countries that impose strict requirements for which organizations should already have systems in place.

Any organization implementing ISO 14001 will have encountered situations where health and safety and environmental issues have merged. Indeed, the importance of health and safety first, and environmental issues second, will have been debated (for example, emergencies or disasters involving fire or major spillage of toxic waste or chemicals) by the organization during the writing of the appropriate procedures.

Recent European legislation has placed even greater strain on businesses, especially SMEs who have found this area of responsibility difficult to mange or have ignored it because they saw it as a high cost burden.

The lack of effective OHS management is, in practice, a high financial cost to industry in any country – not to mention the personal misery caused by such failings. Insurance costs do not always cover the associated liabilities and in the UK non-insured costs are seen as up to 36 times greater than insured costs. The development and acceptance of voluntary standards in the fields of quality, and more recently in the environmental field, has led some organizations to look at the potential benefits of a similar route for managing OHS.

It is with this background in mind that a Technical Committee was set up in the UK, under the auspices of the BSI, to consider whether a voluntary standard would offer any benefits (bearing in mind the plethora of legislation and guidance and some voluntary systems that already existed in the UK). From day one it was accepted that there was no consensus for a certifiable standard in this field as it was argued that all organizations had to have an effective OHS management system in place to meet their legal obligation. It was therefore to be a guidance document and one which would allow integration with ISO 9000 and ISO 14001 (such integration of standards is encouraged in the CIA's Responsible Care Programme – see later in this chapter).

Due to a lack of effective OHS systems and standards, BS 8800 was developed in the UK and is the British Standard for Occupational Health and Safety. It is intended as a guidance document only and not as a basis for certification. (However, at least one certification body has developed a scheme for assessment and certification via the vehicle of ISA 2000 – see Appendix II.)

ISO are currently investigating the need, if any, for the development of voluntary international standards akin to ISO 9000 and ISO 14001 series in the field of OHS management. This is a highly controversial area because health and safety is already highly regulated throughout the world.

The technical committee responsible for formulating BS 8800 were very mindful of the difficulties facing SMEs with so much health and safety legislation and regulation in this field. The integration element was seen to be essential, especially in the light of the Responsible Care Programme (see next section).

The standard itself comprises a main guide which employs a similar approach to ISO 14001, that is, Plan, Do, Check and Act – which is the Deming Principle. Dr Deming is a Quality Assurance 'guru'. Many of the concepts will be familiar to the reader who understands ISO 14001 requirements.

The BS 8800 guide discusses performing an initial review – a Preparatory Safety Review akin to the PER in ISO 14001 – and offers two options to follow when designing the management system. The approach offered in the second option is similar to the ISO 14001 approach (see Figure 5.2) and its structure is in line with other ISO standards.

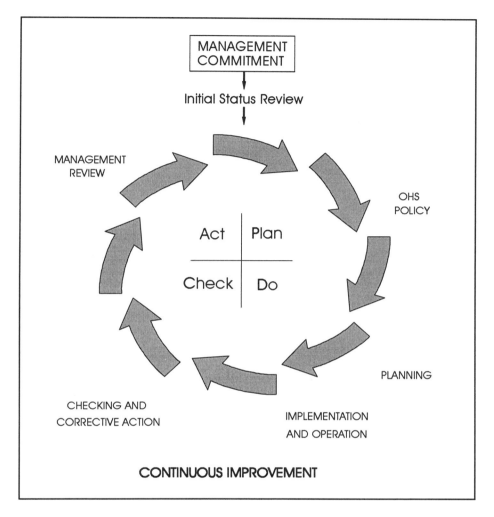

Figure 5.2 BS 8800 implementation cycle

Comparison of BS 8800 and ISO 14001 requirements shows some parallels, or commonalities:

BS 8800	ISO 14001
Initial Safety Review	Preparatory Environmental Review
OHS Policy	Environmental Policy
Planning	
Risk assessment	Risk assessment used in aspects evaluation
Legal requirements	Meeting legal requirements
Safety in product and process design	Environmental considerations at planning stage
Implementation and Operation	
Structure and responsibility	Parallel
Safety training, awareness and competence	Environmental training, awareness and competence
Communication	Parallel
Documentation	Parallel
Control of safety documentation	Control of environmental documentation
Operational controls	Parallel
Emergency preparedness and response	Parallel
Safety of contractors' equipment on site	Conduct of contractors on site
Checking and Corrective Action	
Internal audits – focused upon nature of safety hazard	Focused upon significance of the environmental impact
Accident analysis	Emergency incidents analysis
Management Review	
Verify system meeting objectives	Verify system meeting objectives

From the above comparisons, and by a consideration of Figures 2.1 and 5.2, an organization should be able to begin the integration of these two management systems.

For an organization with no system yet in place, the guide gives a brief outline of how to execute an initial safety review (akin to the PER – suggested by ISO 14001). This advice is supported by six 'how to do it' annexes. The preliminary safety review (PSR) can of course be performed at the same time as the PER. BS 8800 also discusses mechanisms for evaluating probabilities and consequences of accidents. This is risk analysis and certainly there is a counterpart in clause 4.2 of ISO 14001, *Planning*. Thus, in the determination of safety risk, there are many parallels with the determination of environmental risks. The analysis of both could be carried out at the same time.

The implementation of an SMS (Safety Management System) has three phases:

1   A preliminary review of existing safety management

2   Implementation

3   Independent third-party assessment

Having discussed concepts, similarities, differences and how elements of ISO 14001 and BS 8800 can be integrated, the subject of certification to BS 8800 can now be discussed.

ISA 2000

Because BS 8800 was designed as a guidance document only, it was felt that some organizations may still require a standard that they could be audited against by an external certification body. The International Safety Management Organization Ltd (ISMOL) developed a new safety and health management system standard entitled 'ISA 2000 (1996)'. ISA was chosen as the title merely to indicate similarites in structure to ISO standards. ISA 2000 has the capability of being audited against and also integrated with other management systems. The standard has been created with a similar structure to that of other ISO standards, which aids integration. This makes BS 8800 an assessable standard, based on BS 8800 requirements (see Appendix II).

Responsible Care Programme

During the 1970s and early 1980s the chemical industry suffered from a very poor environmental reputation. It was often seen by the public as a major polluter. Its high environmental visibility, coupled with a public perception of 'big uncaring business' resulted in significant pressure from green action groups. The conception and development of 'Responsible Care' was the industry's response.

Responsible Care is the chemical industry's branded program of principles and codes of practice designed to demonstrate that its members take health, safety and environmental issues into account in their everyday operations. Members of the Chemical Industries Association are committed to managing their activities so that they provide protection for the health and safety of employees, customers, the public and the environment.

The program was initially launched in Canada by the Canadian Chemical Producers Association in response to the disaster at the Union Carbides factory in Bhopal, India in 1984. It has since spread to other countries (including the UK) and was launched by the Chemical Industries Association (CIA) in 1989 (the CIA membership representing the major part of the UK chemical industry).

The programme's two main thrusts were:

1   To improve the industry's performance in the areas of environment, health, safety, product safety, distribution and relations with the public.

2   To enable organizations to demonstrate that such improvements were taking place.

The chemical industry is responsible for many environmental issues, of course. Due to the inventiveness of the industry over many years, it has made major impacts on improving the standard of living for billions of people. However, it is also accepted that such improvement in living conditions has associated environmental risks. The chemical industry also believes it has the key to the solution of many environmental problems and is therefore a major player when it comes to addressing sustainable development. The chemical industry saw Responsible Care as helping to regain society's trust in an era of increasing public scrutiny, a crisis of credibility and considerable public mistrust of the industry's activities.

However, there is some public concern about how effectively the CIA ensures that members follow their guidance. In truth, it is rather ineffective. The ultimate threat of expulsion from the CIA is without teeth in an industry where association membership is purely voluntary. All of this undermines the CIA's role as an effective mechanism for self-regulation and suggests that such a programme as 'Responsible Care' may have to be externally certified by an independent body.

This would not be too difficult for an existing certification body who currently certify to ISO 14001, because the common elements within the Responsible Care program can be readily identified, as the following 'Responsible Care Guiding Principles' demonstrate:

1   Organizations should ensure that their health, safety and environment policy reflects their commitment and is clearly seen to be an integral part of their overall business policy.

2   Organizations should ensure that management and employees at all levels, including those in contractual relationships with the Company, are aware of their commitment and are involved in the achievement of their policy objectives.

3   All organizations' activities and operations must be conducted in accordance with relevant statuary obligations. In addition, companies should operate to the best practices of the industry and in accordance with government and association guidance.

4   Organizations will assess the actual and potential impact of their activities and products on the health and safety of employees, customers, the public and their effects on the environment.

5   Organizations will, where appropriate, work closely with the public and statutory bodies in the development and implementation of measures designed to achieve an acceptable high level of health, safety and environmental protection.

6   Organizations will make available to employees, customers, the public and statutory bodies, relevant information about activities that affect health, safety and the environment.

7   Members of the Association recognise that these Principles and activities should continue to be kept under regular review.

Linkages can be seen as follows:

Responsible Care Programme	ISO 14001 clauses
Principle 1) Policy and commitment	4.2
Principle 2) Training	4.4.2 and 4.4.6(c) contractors
Principle 3) Legislation compliance	4.3.2
Principle 4) Environmental impact	4.3.1
Principle 5) Communication	4.4.3
Principle 6) Publicly available communication	4.2(f) and 4.4.3
Principle 7) Review mechanisms	4.6

Therefore, an organization that has implemented a management system to the requirements of the CIA's Responsible Care Programme, has addressed some of the requirements of ISO 14001. To meet all of the requirements of ISO 14001 may only require a few additional procedures and for such procedures to be adequately documented.

In summary, any company that has complied with the CIA's guidelines and now seeks external certification – for the reasons of credibility given above – would have little difficulty in achieving such independent certification to ISO 14001.

BS 7799:1995 British Standard Code of Practice
for Information Security Management
Information is the key to any successful business. The more correct and up to date the information is, the more often the organization can act in the most appropriate way to ensure its financial success. This information can be gained, analysed and stored in a variety of forms, such as computerized data, documents, tape recordings etc.

Appropriate security should be applied to all of these media, whilst ensuring easy access for the right people at the right time.

Organizations now face increasing security threats from a wide range of sources. Information Technology (IT) systems and networks may be the target of a range of serious threats including computer-based fraud, sabotage, vandalism, computer viruses and hackers. Such threats are expected to become more widespread and increasingly sophisticated in the future. System failures and disaster scenarios must also be considered as a risk element. Against this background and following requests from industry in the UK, BS 7799 was written. This was written as an auditable standard which, if implemented within an organization, would reduce the exposure to the risks outlined above.

BS 7799 is a set of guidelines and recommendations which certification bodies can interpret for the needs of any organization or company. BS 7799 is a code of practice, not a rigid specification. Adaptability is inherent to the code's structure. Preparation for compliance with the Standard can be broken down into three phases:

Phase 1) Preparatory security review. Each area of activity is analysed and those where information security is most vulnerable identified.

Phase 2) System implementation – to be fully documented.

Phase 3) Assessment and certification ensures that the system complies to BS 7799 and that the extent of the information security system (that is, its breadth, depth and width) accurately mirrors what senior management have documented.

Although the Standard describes 109 controls, only 10 are considered 'key'. These key controls are essential for all organizations, whereas the others can be selected, depending upon the nature of the business.

The key controls are listed below, alongside the equivalent ISO 14001 clauses:

BS 7799 Key Controls	ISO 14001 Clauses
1) Security policy	Environmental policy
2) Security organization	Organization and responsibility
3) Information security education and training	Training, awareness and competence
4) Reporting of security incidents	Nonconformance and corrective and preventive actions
5) Virus controls	Operational controls
6) Business continuity planning process	Emergency response
7) Control of proprietary software copying	Operational controls
8) Safeguarding of organizational records	Control of environmental records
9) Data protection	Operational controls
10) Compliance with security policy	Compliance with environmental policy

Figure 5.3 illustrates the BS 7799 implementation cycle.

Although some of the above do not have identical counterparts in ISO 14001, there are several areas of obvious commonality. An integrated system would cover both – for example the use of risk assessment techniques (as follows).

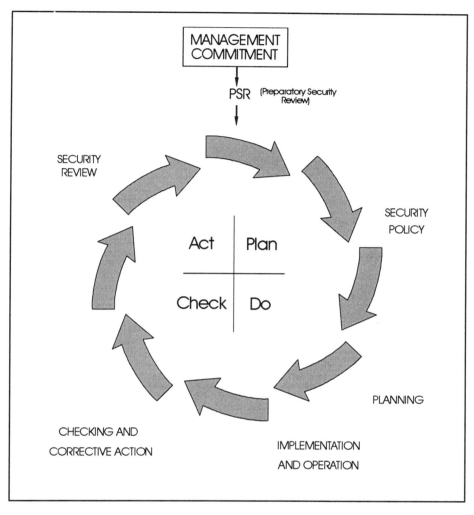

Figure 5.3 BS 7799 implementation cycle

As was discussed in Chapter 3, when evaluating environmental aspects for significance, risk has two components: probability and consequence. In the case of security management these translate to:

1   The realistic likelihood of such a breach of security of data occurring in the light of prevailing threats and controls.

2   The business harm likely to result from a significant breach of security, taking account of the potential consequences of failures of information confidentiality, integrity and availability.

The code of practice expects an organization to take account, where relevant, of existing legislation. In the UK this is the Data Protection Act, the Computer Misuse Act and relevant legislation relating to software copyright. Thus in this respect there are further parallels with ISO 14001.

Forestry Stewardship Council (FSC certification)

It is widely accepted that forest resources and associated lands should be managed to meet the social, economic, ecological, cultural and spiritual needs of present and future generations. Furthermore, growing public awareness of forest destruction and degradation has led consumers to demand that their purchases of wood and other forest products will not contribute to this destruction, but will rather help to secure forest resources for the future. In response to these demands, certification and self-certification programmes for wood products have proliferated in the marketplace.

So states the introduction to the FSC principles and criteria for forest management. *Self-certification* is the process whereby an organization states that their forest management techniques ensure sustainability. Such self-certification is becoming a less attractive option for forestry organizations as, increasingly, customers require independent third-party certification.

The Forest Stewardship Council (FSC) is an international body which accredits certification organizations in order to guarantee the authenticity of their claims. The goal of the FSC is to promote environmentally responsible, socially beneficial and economically viable management of the world's forests by establishing a world-wide standard of recognized and respected *Principles of Forest Management.*

These ten 'Principles' are listed below alongside applicable ISO 14001 elements:

FSC Principles	ISO 14001 elements
1) Compliance with laws and FSC principles	Compliance with legal and other requirements – clause 4.3.2
2) Tenure and use rights and responsibilities	No strong linkage
3) Indigenous peoples' rights	No linkage
4) Community relations and workers' rights	Communications – links to clause 4.4.3
5) Benefits from the forest	Reduction of waste – commitment to continual improvement equates to clause 4.2
6) Environmental impact	Clause 4.3.1
7) Management plan	Clause 4.3.4
8) Monitoring and assessment	Clauses 4.5.1 and 4.4.2
9) Maintenance of natural forests	No strong comparison although operational controls – clause 4.4.6 applies
10) Plantations	As above

Looking at these ten principles, there are some that have an equivalence with ISO 14001 clauses. However, there are some differences in detail. The main elements that appear to be missing from the FSC are the ISO 14001 requirements for a documented system and a requirement for management system audits. The 'sociological' Principles 2, 3 and elements of 4, address the needs of minority groups – local communities and indigenous peoples – and do not have a direct link to ISO 14001 requirements. Although there could be arguments put forward that, by minimizing waste and protecting the environment, there is by definition an opportunity for improvement in social conditions of such groups.

Thus the two standards are capable of being audited at much the same time, with a large proportion capable of being verified by only one 'check list'. Dual certification to ISO 14001 and the FSC is already happening – teams of auditors review the common elements of the standards, whilst technical specialists audit the specific ISO 14001 or FSC requirements.

The process of integration

This chapter has shown that, in principle, the integration of environmental systems with other systems is possible because of their many similarities. Although there are many similarities between the Standards discussed – both in specific requirements and underlying philosophy – there are differences. However, these differences are in detailed requirements only, and have been indicated.

Again, it is not the purpose in this book, whose focus is ISO 14001 implementation, to describe in detail how to integrate two or more management systems. The purpose is to highlight the concepts and, due to the complexity of such an undertaking, each organization will have to derive its own plan of integration. The essential steps of integration are described below.

A good starting point for the integration of management systems is to perform a Preparatory Integration Review (PIR). This is similar to a PER from ISO 14001 but has a much wider scope. The PIR needs to be performed on a comprehensive review of the business in terms of the way it manages quality, health and safety and the environment. This review should be carried out at executive level as a high level of commitment will be required.

Analysis of the business processes

A typical PIR would cover:

Identification of:

- ¤   Duplication, overlaps and gaps in systems

- ¤   Legislation requirements

- ¤   Stakeholder requirements

- ¤   Standards of operation – ISO 9000 and others

- ¤   Costs of each process

- ¤   Hazards and risks

Formulate opportunities and policies:

- ¤ To rationalize policies and procedures

- ¤ To redefine responsibilities and accountability

- ¤ To reduce costs

- ¤ To reduce risks

- ¤ For common element integration

Plan:

- ¤ Identify key personnel (champions) for implementation

- ¤ Establish objectives and targets

- ¤ Prioritize and plan implementation

- ¤ Establish training and awareness needs

Do:

- ¤ Communicate policy objectives and targets internally

- ¤ Implement training plan

- ¤ Implement costs and improve actions

- ¤ Establish methods for performance indications

Check:

- ¤ By auditing to monitor performance

- ¤ By auditing to identify opportunities for improvement

Act:

- ¤ To implement and evaluate improvements

- ¤ To analyse root cause to prevent recurrence

Certainly, a PER, a PIR, and a PSR can all take place concurrently. Integration of quality assurance, environmental management and occupational health and safety within an organization is, without a doubt, a part of the business plan in the majority of organizations who have combined two or more systems.

Considering all these standards, there are many common clauses that need to be satisfied. It makes sense not to duplicate them. For an organization that has two systems running side by side, integration can be on several levels:

1   **Documentation systems can be combined.** Purely by combining the documented systems and considering the hierarchy of documentation – the 'pyramid' of documentation (see Figure 3.5) – it will be seen that the lower level documents overlap first. Combining such documents can be a relatively easy task. The same 'combining' applies equally to quality, environmental and health and safety records.

2   **Internal audits and management reviews can be held concurrently.** Training needs analysis can be addressed concurrently.

3   **Policies can be combined.** At a higher level of operation (the apex of the pyramid) are the separate policies. To combine these into one meaningful policy requires much more thought in order to produce an integrated policy that does not require frequent updating, for example.

Consider a company that manufactures plastics containers for the food industry. At an operational level, for example, there may be a procedure with detailed work instructions for the screen printing of plastics components:

*   **Quality Assurance** would need a procedure that focused on the requirements for the operator when mixing inks, preparing screens, degreasing plastic components, etc. in order to ensure the quality demanded by the customer (such as the colour of a logo or the clarity of text).

*   **Occupational Health and Safety** requirements would dictate that when such inks etc. are in the process of preparation, an instruction would be available describing the safety precautions to be followed (for example, good ventilation and protection of the skin by barrier creams). A requirement for hygienic manufacturing conditions may be dictated by the customer – as the containers are for the food industry.

*   **Environmental Management** controls would require that the operator controlled the disposal of any waste solvents or inks as well as knowing what action to take in the event of accidental spillage.

Three work instructions would be required: one each describing the Quality Assurance, Occupational Health and Safety, and Environmental Management System controls respectively. However, it would be relatively easy to write only one work instruction emphasizing the three 'types' of control required.

One approach adopted by several ISO 14001 certified organizations that have had quality assurance procedures in place for many years, is to control and review the integrated procedures under the banner of the quality system. Such procedures are a logical extension to their Quality Assurance system – the focus, however, is on Health, Safety and Environmental issues. This approach appears to work very well for them.

Attitudes to integration

An organization may decide that the integration of its separate certified management systems is necessary from both a use of resources and costs point of view. It makes sense therefore to understand the stance, expectations or opinions of the various interested parties.

The ISO 14001 Standard

The introduction in the ISO 14001 Standard states that ISO 14001 is not intended to address aspects of occupational health and safety management. However, the Standard does not discourage an organization from developing integration of such management system elements. The annex to the Standard briefly mentions that integration of environmental matters with the overall management system can contribute to the effective implementation of the environmental management system as well as to efficiency and to clarity of personnel roles.

The certification bodies

Integration is encouraged by certification bodies, as it reduces certification costs (due to combined audits) and less duplication of documentation for the client.

However, the practicability of simultaneous certification is dependent upon the correct mix of auditors being available. There are logistical problems getting the right mix of auditor skills together at the right time and place. The development of the multi-disciplined or 'hybrid auditor' (see Chapter 7) may ease this practical limitation in the longer term.

The accreditation bodies

Accreditation bodies encourage integration of management systems. This is clearly stated within the introduction to Accreditation Criteria (see Appendix IV). However, the reference also makes clear that the initial assessment (meaning the stages of pre-audit, desk-top study and certification audit) should be performed as a separate exercise from any other certification.

Internal and external auditors

The internal auditor

For the internal auditor the requirement to audit an integrated management system should not be too onerous. Guidance could be given by the quality, health and safety and environmental 'experts' respectively in house to the auditor via briefing or guidance notes and perhaps pre-prepared checklists. Usually, internal auditors are experienced individuals from within the organization and a program of training may be sufficient to raise their level of awareness of environmental, health and safety and quality issues.

The external auditor

The external auditor representing the certification body may find an integrated system more of a challenge. At the time of the ISO 14001 certification audit, it may be that an organization requires a concurrent ISO 9000 assessment. The challenge here is one of planning and preparation. The team leader can organize the team members to audit different requirements with pre-prepared check-lists and spend some time during the audit monitoring progress of both 'disciplines'. A greater challenge might be presented during routine surveillance visits when only one auditor might attend. The auditor will need to become very familiar with two or more management systems in a short space of time, to enable a meaningful audit to take place and also to add value to the client's system. This challenge may be made harder if the auditor has not been to the organization previously – either at the certification audit or a previous surveillance visit.

Summary

This chapter has examined the current management standards that are being implemented in several sectors of business. The standards covered are not exhaustive. There are others concerning training programmes and 'moral' responsibility in sourcing materials. However, the standards that are the focus of this chapter are the most widely used by the business community, and there are many similarities in their requirements and structure:

- The initial status review

- Setting out policy and objectives

- The manner in which organization is defined

- The planning and implementation of the system

- Measuring performance – checking and reviewing

This is all based upon the Deming cycle of 'Plan, Do, Check and Act'.

Clients with several management systems, each of them operating independently, are increasingly asking themselves whether integration would save them money and, indeed, when it comes to certification, whether the certification bodies give more value by combining assessments and surveillance activities.

In fact, certified organizations are now seriously considering the concept of 'value for money' in the area of certification. Whilst holding separate certifications, either from one or several certification bodies, the customer expectations of the future will be 'one certification body and one integrated certified management system covering all aspects of the company's management activities.'

The ISO Technical Management Board are investigating the possibility of producing an integrated management system standard which will effectively incorporate existing standards on health, safety, quality and environment. This standard will lead to the development of a framework standard that will encompass generic management system principles. This would enable further development of guidelines for all types of management systems as briefly covered in this chapter.

Chapter 6

Case studies

This chapter consists of the case studies of four organizations who have successfully implemented ISO 14001. The four organizations are very different in terms of the industry sector they operate in – medical devices, metal wire, electric light (lamps) and air conditioner production. The style and content of each of the case studies varies. This is not only a reflection of different authors' styles but also the fact that each of these organizations faced their own particular challenges during implementation.

There is, however, a similar structure for each case study, comprising: an introduction to the author; an introduction to organization and why the environmental management system was implemented; and a focus on the challenges and issues faced by the organization.

It is also worth noting that the 'certification routes' were different in these case studies. Only one was certified directly to ISO 14001. Two of the others converted their BS 7750 certification to ISO 14001, whilst the remaining one obtained both BS 7750 and DIS/ISO 14001, being the (then) draft ISO 14001 standard.

Seton Healthcare Group plc

Written by Neil Stewart, advisor to Seton Healthcare Group plc., in consultation with the Technical Director, Graham Collyer and Environment Manager, Ingrid Osterburg. 3rd December 1996

The Company

Seton Healthcare Group plc (referred to as 'the Company' in this text) is a fast growing entrepreneurial company based in northern England. It manufactures a range of surgical and medical bandages, dressings, orthopaedic soft goods, prescription and 'over the counter' pharmaceuticals, continence care, and other healthcare products as well as sport and leisure goods. These are marketed and sold both in the UK and for export.

The Company started life in the 1950s as a private company and after rapid growth floated as a public company in 1990. Its turnover in 1995/96 was £84.9m and profit before taxation was £16.4m. It employs around 1,000 people on five sites in the UK.

It elected to be one of the first companies to apply for certification to the new Environmental Management Standard BS7750. During 1992 and 1993 the Company took a leading role in the pilot study in the textile sector. A small team, referred to throughout the text as 'the Team', was assembled under the leadership of the Technical Director to oversee the implementation of this project. Certification on its headquarters sites was eventually achieved in January 1996. The Team consisted of an Environment Manager, an advisor and some clerical support – all of whom were assigned part-time to the project. The Environment Manager spends 50% of her time on environment and 50% on quality management.

During a routine surveillance visit by the certification body, there was a seamless transition from the existing BS 7750 certification to ISO 14001:1996.

Background

In comparison with other industrial companies, the Company is not a major polluter. The pharmaceutical products which it sells are made from existing and known active agents and the Company formulates these into effective medicines. The Company was originally a textile concern and its entry into the healthcare market was brought about by a very successful, textile niche product

– Tubigrip – which is an elasticated bandage. It does not wash this product or treat it in any way during production that might create quantities of waste water and chemicals. The process that has the biggest direct impact on the environment is a coating process that uses solvent-based adhesives.

This background illustrates that the Company was not motivated by any perception of direct environmental liability when it set out to achieve certification to BS 7750. The Company is proud of its brand name as a symbol for healthy living and, being market-led, is keen to participate in any activities which its customers would regard favourably. The Company knew at the outset from its market research that nurses, an important group in determining the choice of its products, are concerned about the environment. It was also aware that one of its most important customers in the retail market is keen that its suppliers are environmentally responsible, and carries out supplier audits from time to time.

The Team had a priority task. This was to understand the benefits the Company would gain from having an environmental management system. These potential benefits were then presented to the Board of Directors who issued a statement describing the benefits as follows:

i)   An external statement to our customers and other stakeholders about our ethical intentions will strengthen our brands.

ii) A properly constructed environmental management system will allow us to anticipate real business risks in the market place.

iii) There is evidence of increasing interest in environmental issues by investors whose interests are fundamental to our business success.

iv) Achieving certification is consistent with our belief in being at the leading edge, where this can be done without too high a risk (risk being in terms of financial loss).

Thus the benefits which the Company set out to obtain are strongly related to market and business excellence. This is fully consistent with the culture of the Company and also reflects the strong sense of community involvement which was always a high priority of its original private owners.

During the course of establishing the new management system, six 'critical issues' emerged. These were:

1   Getting and maintaining the commitment of the Board and other key managers in the Company to the project

2   Developing an effective methodology to evaluate environmental effects

3   Integrating the project into the existing working practices of the Company

4   Understanding and managing the scope of the system, including the extent to which suppliers and customers would need to be involved

5   Developing an effective training strategy at different levels

6   Maintaining the momentum of the project after certification is achieved on the first sites

Critical issues

Commitment
In a modern European company, whatever its corporate culture, it is difficult to achieve any programme of change without the full and active involvement and agreement of the workforce (at all levels). It was the Team's guiding light from the start that the environmental programme was not something to be tucked away within the Technical Department but something owned and managed by the Company in total.

For this reason, one of the Team's first steps was to go and interview some managers: firstly, to judge their reaction to a programme based on environmental excellence; secondly, to get a view from them about what they might expect such a programme to do for them in their management roles. Without exception, these initial interviews were positive. The majority of managers were not only in favour of taking an environmental leadership position but also expected positive benefits if this could be done with acceptable use of costs and resources.

In practice, commitment was achieved within the Company by acting at three levels – the Board, the Managers and the Workforce.

Within the Company, as elsewhere, no major project can be effective unless it has a clear-cut business rationale. The Team took the view that environmental management had to be part of the Company's corporate objectives and that the environmental programme had therefore to be seen as a business project, subject to commercial scrutiny in the conventional way. The Team assembled a project statement focused on the key objective of attaining certification and laid out reasons why this seemed to be a sensible business aim.

The practice of reporting to the Board was continued annually. All statements included requests that the Board underwrite the effort required to achieve environmental objectives. By this means, the Board was kept fully informed of progress. Furthermore, in each of the later years, the Team produced a leaflet laying out the Company's policy, objectives and targets for distribution at the Company's Annual Meeting. The intention was that shareholders (and other stakeholders) would also be quite clear about what the Company was trying to do. Throughout the process and up to the present day, the Board has been very supportive of the environmental programme.

Because the Team believes that it is essential to involve all the Company in the environmental programme and not just a small team in the Technical Department, formation of the Environmental Management Review was a necessary early task. From the start, the managers and directors who were jointly responsible for the areas to be covered by the environmental programme were brought together to plan the details of the project. The Team saw itself as offering a service to these managers, rather than as directing activities. The process of biannual management review has continued and is an essential part of the programme.

The Company had also achieved certification within the 'Investors in People' (IIP) programme, which is a training and consultation framework. Part of this programme requires that members of the workforce are briefed on important business issues several times a year. The Team therefore developed materials to be used in Team Briefings. They put together a small leaflet in the early part of the environmental programme to alert the workforce and ask for feedback, and later repeated the process with a leaflet explaining the background to the Company's Policy, Objectives and Targets. Materials for noticeboards and informal presentations were also assembled as part of this process of briefing. Once again, the participation of the workforce was positive and some good suggestions emerged, notably related to packaging and waste. The comments which arose from this consultative process were included in the Preparatory Environmental Review.

Environmental impacts

As the Company was one of the first to be involved in developing an environmental management system, techniques for carrying out environmental effects evaluation were not available 'off the shelf'. The Team was clear that a methodology had to be developed for assessing environmental risks which was:

a)   Clear, logical and self-consistent

b)   Defensible from criticism by auditors and other external bodies

c)   Based on the Company's own assessments but, in turn, dependent on its own demonstrable competence to judge environmental risk

This latter point meant that the Team had to develop the capability to understand and judge environmental effects in-house. Even where specialist external services were used, the Company had to have the capability to understand and interpret for action, any advice received from external third parties.

In the early days of the project, three sources of advice were invaluable in helping to develop what the Team believes is very sound practice in evaluating environmental effects. The sources of advice were:

The network of BS 7750 pilot scheme members

The Company was a leading member of the BS 7750 pilot project in the textile sector; this pilot project also included members of trade associations, academic specialists and members of supplier industries (notably chemicals). This provided a very valuable network for comparing and contrasting experiences and developing methodologies.

The Certification Body

The initial visit from the Certification Body that was eventually to audit the management system took place at a time when it too was learning. Thus the Team was able to check its ideas against those of the external auditors and this proved to be a very useful learning process. This learning process took place at the time when verification organizations were involved in their own process of developing guidelines for the interpretation of the new management standard under the leadership of the then NACCB (now UKAS).

External training

Finally, as the Company had appointed an Environment Manager (who also had duties in the quality management systems of the Company), this person participated in relevant training programmes, which proved very valuable. The programme of environmental assessment provided to standards set by EARA (Environmental Auditors Registration Association) was especially valuable.

With all this developing experience, the Team was able to assemble a working process of assessing environmental effects, which is essentially:

- Listing environmental aspects

- Understanding their potential environmental impacts

- Understanding the likelihood of environmental damage being caused

- Understanding the consequences of such environmental damage

- Evaluating impacts on a scale of significance

- Assessing the control and minimization procedures which are required and which may or may not be in place

- Understanding of the possibility for improvements which can contribute to the Company's objectives and targets

Having established a methodology for carrying out environmental effects evaluation, the Team was then confronted by the problem of 'how to get started'. The approach to Preparatory Environmental Review was based on three factors:

i) What the managers considered to be important in the initial interviews.

ii) What the workforce considered to be important in the initial consultation sessions.

iii) What an 'environmental pressure group', such as Greenpeace, would think of the Seton Healthcare site.

This last question was prompted by the Team performing a 'walk around' of the site when assembling the initial list of environmental aspects.

Once the Team had assembled the initial Effects Register in this way, an audit process was established to refine and update the Register on a continuing basis. This audit process was designed to explore environmental effects in detail. It was scheduled to reflect a developing understanding of where the most significant effects were to be found.

Environmental audits were carried out by scrutinizing the activities in specific departments. Additionally, the audit programme included product group life cycle analysis.

Prioritization was determined either by the need to understand more about an environmental impact or by the perceived severity of the imports themselves. The Product Group studies included an appreciation of the impact of both suppliers and customers. Systems Audits (see below) were also included in this programme. The result has been to build an increasingly detailed and precise understanding of the Company's environmental impacts and how they are managed.

Integration

The Company had been one of the first in its sector to be awarded a BS 5750 Part 1 (ISO 9001) Certificate in the 1980s. It was therefore logical to build its BS 7750 system around its existing quality management processes. The existence of the quality systems, nevertheless, proved to have advantages and disadvantages. On the positive side, the Company was well used to writing effective work procedures and the workforce was well used to being audited and seeing this as an effective and helpful management tool. Thus Systems Audits to check the environmental management system could be incorporated into the parallel activities for quality management. Management control activities such as management review and policy formulation, and the continual improvement ethos, fit quite comfortably into both types of system.

Nevertheless, there are some important distinctions between quality and environmental management which determine the way in which the respective systems should be established. At the core of a quality system is an understanding of the needs of customers; the procedures within the system are designed to fulfil customer needs. At the core of an environmental management system is an understanding of environmental effects and the control of these together with progressive improvements. In some ways, environmental management systems, with their focus on control, are very similar to health and safety management. The Team was particularly encouraged to view BS 7750 in terms of controlling technology, via procedures and operator skills, by the certification body with whom it was dealing and this had implications which will be noted below.

The most important milestone in integrating the environmental system within the working practices of the Company, however, occurred when the Team came to propose objectives and targets to the Board. The Team was able to isolate a list of environmental effects where improvements were desirable, feasible and sensible and formulate these into specific improvement projects. Although some of these environmental projects were led by the Environment Manager herself, many of them were run by the line managers most directly involved with the processes that created the effect. Progress was reported in the regular Environmental Management Review meetings. In this way, a big step towards management integration had taken place. For example, Marketing and Purchasing Directors took the lead on the packaging project, the Site Director took the lead in monitoring energy efficiency and the Fleet Administrator set targets for fuel consumption and other aspects of the Company's motor and truck fleet.

In some senses, this is inconvenient to the Team members who sometimes feel accountable for something where the control lies elsewhere. Nevertheless, in order to achieve an integrated environmental programme, it is essential to have line management fully involved and leading aspects of the programme.

Scope

It has been pointed out that the Company is not a major polluter on the scale of, say, the chemical industry. Even though the Company is part of the textile industry (because of its involvement in textile dressings), it does not engage in wet processing and therefore does not have any major use of water or potential for polluting water courses. The most environmentally risky process in which it is involved relates to the use of adhesives which contain solvents. For this reason, the preoccupation of the Company from the start was as much with indirect effects as with direct effects. There are a number of issues here, such as:

i) The Company is a major provider of products to the hospital sector which disposes of its waste products by incineration. Thus the Company needed to be aware of any requirements which its products might have to meet from that source. This was particularly the case as much of the preparatory work on the Environmental Management System took place during the period when the National Health Service was adjusting to its loss of Crown Immunity and was therefore exposed to new and very demanding regulatory requirements for its incineration operations.

ii) The Company produces foam-backed bandages where polyurethane foam is applied with an adhesive to a textile product. Both the foam and the adhesive are purchased from third parties. This raises a number of indirect environmental issues, such as: what solvents are used in the adhesives; and by what processes is the foam manufactured.

The first thought of the Team was that it should carry out some form of evaluation of its suppliers. Accordingly, and in common with many other companies, the Team devised a questionnaire with the help of the Purchasing Department and sent this to a selection of its leading suppliers. The questionnaire asked a number of questions about environmental management systems as well as about the specific products being purchased.

This questionnaire still exists but its use has been reduced considerably. Completed questionnaires were received and as a statement of intent to our suppliers, it did some good. However, it probably also did some harm because it had a very broad scope and must have been regarded as well-intentioned but useless by some of its more discerning recipients.

The Team realized that the supply chain involving suppliers, Seton Healthcare Group plc and customers needed to be more fully understood by themselves. This meant carrying out life cycle assessment studies of some of the key products sold by the Company.

This in turn gave some problems. Life Cycle Assessment (LCA – see Appendix I) is a difficult, complex and developing area and the Team was initially taken aback at the amount of work it had seen in some examples. They were unsure where the LCA started and finished. This was not helped by the fact that even the most complex LCAs seemed to be controversial. The Team discovered that its protocol for environmental effects evaluation could easily be extended, with appropriate judgement, to form the basis for a Life Cycle Study or LCS (an LCS being a much more easily managed option than an LCA). The most important supply chains of the Company have been chosen partly because of their size in its portfolio and partly because of an initial perception of environmental impact. These supply chains are now being subjected to this LCS approach. The Com-

key word search ISO/YCOP

— 337 · 7 BAT

— 658 ·A8SHH1

pany can easily show that it bases its understanding of its indirect effects on a rational and soundly-based methodology.

In parallel with this belief that the most important way to tackle indirect environmental effects is to understand and influence the supply chains of which it is a part, the Environment Manager has taken a leading role in the environmental groups set up by the relevant trade associations. Within many industries, trade association are taking a leading role on environmental issues.

The 'scope' of an environmental management system can be the geographical location of the organization but it is usuaslly described in terms of the organization's activities. At the outset, the Team and the Board decided that to cover all the manufacturing sites would be too large a task, particularly as the Company is a fast growing, acquisitive one with frequent additions to its product portfolio and changing use at manufacturing sites. For this reason, the scope of the system was initially set as the headquarters sites, which included mainly the textile operations together with the group's Marketing, Purchasing, Technical, Distribution and main management services operations. This was the scope of the initial certification and the Company now has specific plans to extend its management system throughout the group.

Training

There is an important distinction between *awareness-raising* activities discussed under *Commitment* above, which serves the purpose of alerting the workforce to what is happening and why, and *training* which is designed to change behaviour. As already noted, the Team Briefings within IIP gave a useful vehicle for awareness-raising. The Team was also aware of the need for training at three levels:

i) **Environmental specialists**: Training of specialists in environmental risk assessment and the related techniques.

ii) **Managerial and supervisory**: Training of those managers who are responsible for setting specifications for the Company's products or products and services purchased by it, so that they are aware of the environmental implications of their activities.

iii) **The workforce**: Broad training of the workforce, especially focused on the environmental effects in the areas in which they are active.

The first level of training was achieved quite easily by selected members of the Team going on appropriate and recognized courses. The Environment Manager had no previous qualifications in this area but, with her quality and technical background, she found the courses easy to absorb and useful. One advisor, who has a background in synthetic chemicals and an environmental sciences degree, has also been on an EARA-recognized course.

It was much more difficult to put together an effective training strategy for the other managers and the workforce and to devise a system whereby this training would be updated (as employees might change jobs or job specifications might change). The training must be owned by line management, although the Team have created a training module to be used when managers see the need.

In practice, the Team saw the need to be very persistent in saying that environmental care is part of everyone's responsibilities. They have therefore set out guidelines for environmental competencies within 'Job Descriptions'. By this means, each individual is invited to explore this aspect of performance as part of the annual appraisal system. Training needs at individual level are then flagged up for attention. As this is a formal critical procedure within the environmental system, its effectiveness is checked and maintained by regular audit.

Momentum

It is quite often the case, in an organization with highly-motivated and enthusiastic staff, that new achievements are being sought all the time. Seton Healthcare Group plc was no different. All the managers were agreed that obtaining certification to BS 7750 was only the first step on a long journey to environmental and business excellence. However, other pressures also crowd in. In practice, the Team was confronted with the problem of how to sustain the momentum which it had created and retain the interest of managers and workforce.

This issue is currently being addressed. It helps considerably that some 'deliverables' already exist. Reductions in the company's direct and indirect environmental impacts have been demonstrated. These improvements include some real cost benefits and product innovations. The management structure, integrated into normal work practice, which has been established as part of the process of certification, will be essential in driving the programme forward. There is no lack of enthusiastic people on the new sites waiting to be drawn into the overall process. Thus the momentum can be self-sustaining to some extent.

However, there is one additional point. It seems to the Team that maintaining our progress with the new system will depend on maintaining a clear sense of benefit in the minds of both the Board and the managers who are involved. There is a danger in taking a 'health an safety management' approach to environment – health and safety are real issues which impact on people's everyday life, but environment is rather more abstract and distant. It requires a different type of commitment.

Thus the Team sees the need to revisit the strategic basis on which the environmental management system was established in the first place and make sure that all the managers and directors involved can see quite clearly the benefits to themselves. Environmental management has to be seen as a strategic issue (that is, encompassing longer term goals and considering wider issues of 'sustainability'), setting a context for such practical systems as ISO 14001. This becomes critical once the sense of excitement at achieving certification is passed.

Tinsley Wire Ltd

Written by Mike Osborne, Environmental Manager, Tinsley Wire (Sheffield) Ltd
January 1997

Introduction

The company processes around 120,000 tonnes of mild steel wire products per annum on its 35-acre site in the Lower Don Valley, in South Yorkshire, UK. Numerous chemical substances are used for pickling, electroplating and hot-dip galvanizing processes.

The area around the factory was heavily industrialized until the mid 1980s and since then has undergone significant change. The surroundings have now become one mainly concerned with retail and leisure pursuits. In addition to this, a pre-existing freight railway line which literally runs through the centre of our 'figure eight'-shaped site was converted to use for the new 'Super-Tram'. A canal/towpath which runs alongside the tram has been declared an environmental corridor by the local development association. The site has a long boundary, virtually all of which is open to public view. A retail park has taken over a site previously occupied by a scrap merchant's yard, and this is situated alongside our storage facility for 36% hydrochloric acid, pickling plant and effluent treatment plant. With all of these developments replacing traditional industry, the perceived 'image' of the area has undergone significant change.

With our company being one of few remaining in production in this area, it was realized that we needed some system to control environmental issues arising from the company's operations in order to achieve long-term security and acceptance. Around this time, BS 7750 was first published and the company participated in the pilot application of the standard. Particularly difficult issues surrounded the development of an Environmental Effects Register and how far we should be looking at indirect effects. This was initially complicated to some extent by the appearance of ISO 14001, but the decision was eventually made to base the management system on the ISO Standard.

Commitment

Initial commitment of the management team towards environmental issues is of paramount importance. There are many difficult decisions to be made along the 'environmental' road. In our company, management of both environmental and health and safety issues are seen as being of paramount importance to its future and success. In order to gain this commitment a Steering Committee was formed to discuss and implement policy and procedures. This meets at approximately six-weekly intervals for around two hours and consists of the Managing Director, Directors of Production, Engineering, Human Resources, Sales, Finance, the Environmental Manager, Safety Manager and Personnel Manager. Particular issues on the meeting's agenda include any item of noncompliance, laws, regulations, results of audits, actions arising from the Management Programme etc.

In addition, the Environmental Manager prepares a summary report for discussion at each meeting of the company's executive committee. This helps to raise the profile of the environmental management system, keep the management team aware of process improvements, and 'prepare the ground' for changes that may be necessary.

Environmental aspects

This was the one area that initially gave rise to some confusion and great difficulty. It was easy to fall into the trap of 'knowing from experience' where the problems lay and starting to bring about change prior to adopting a well thought out systematic approach, 'second time around'. This happened largely because the company started to try and recognize where it's environmental commitments lay before publication of BS 7750.

It soon became obvious however, that in order to prevent an ad hoc improvement plan which placed scarce resources with inappropriate projects, a systematic evaluation was needed.

This evaluation was relatively easy in some cases. A visible plume of smoke from a chimney was an obvious environmental aspect. Developing an improvement plan and monitoring progress was straightforward (that is, a plume could be seen to be less visible).

However, a different approach was needed when, at a first glance, processes did not show any obvious problem areas involving emissions to either atmosphere, water course or land. There were, nevertheless, potential areas where improvements could be made: such as reducing the risk of accidental emissions to atmosphere or spillages to water and land.

As a company, we had already evaluated several Health and Safety issues relevant to our workforce using a simple risk assessment methodology, ranking risk in a low, medium or high category. Following the success of this type of system in revealing some significant risks of a health and safety nature, it was decided to follow a similar approach for environmental issues.

After a full investigation into what environmental legislation was relevant to the business, the processes in each of our production, engineering and technical areas was broken down into separate line diagrams (showing inputs and outputs from the process being carried out). Each of these diagrams were then assessed for potential environmental impact to environmental media (that is air, water, land) based on the consequences of such an event and the likelihood of it happening. Past experience was found to be of great value when making such assessments. Comparing criteria set within the legislation for emissions to air, water and land with actual monitoring data allowed further objectives to be set – objectives which may not have been apparent initially.

By using a simple point-scoring system, each considered aspect was placed into a low, medium or high-risk category. The number of points awarded to each is quite arbitrary; the important thing is to associate a high number of points with a higher degree of risk. (Commercial software can be purchased, if desired, for this type of assessment.)

The higher-risk aspects can then be the focus of a management programme. The associated objectives arising from the plan can be used to define individual personal targets within the organization. For example, it may be found that there is a potential for some parameter in a discharge consent for liquid effluent to be compromised by a certain chain of events within the factory. The objective arising from this discovery may be to achieve 100% compliance with the discharge consent.

Personal targets arising from this may involve individuals with differing action points such as:

- Compiling written procedures

- Changing a substance currently in use

- Amending a production process

- Using a monitoring programme

- Developing a recovery plan should an accident occur

Each would have a specific quantifiable target in terms of time for completion or a measurable parameter (such as rate of use).

By repeating the same process year on year, the overall degree of risk to either the environment or the business will be found to be reduced. Therefore, as time passes, the medium and lower risk issues not initially considered will slowly be covered by 'the system'.

As noted above, in the early stages of implementing a management system, the initial approach may well be to think the important issues are already known. This impulse should be controlled – it can lead to some issues not being fully explored or even discovered at all. Another danger in the early stages of implementing a management system is to try and commit the business to too many objectives, spreading valuable scarce resources too thinly to be fully effective.

Another point to consider is that, if researched fully, an environmental objective will almost certainly be accompanied by a good business objective. As an example, metal process scrap is a costly enough waste of raw materials. However, the disposal cost of this scrap material is an additional expense that has to be borne by the organization.

The environmental management system was developed to recognize environmental based legislation which was of immediate relevance to our business. I say 'of immediate relevance' because over the past few years, I have seen other environmental management systems which have generated much paperwork, and have occupied much management time, in keeping up to date with all legislation. Much of this was only tenuously relevant. This makes the system difficult to compile in the first instance and almost impossible to update and work with. We only have legislation in this register which has direct impact on the company's day-by-day operations. The register is still reasonably 'weighty' but being directly relevant makes it a very useful working tool that is referred to time and time again. There are many useful publications, both in printed and electronic form, to help with this task, many of which offer updating services. We also find abstracting services from Trade Associations and other companies who maintain databases of legislation (with updates available by subscription), an ongoing source of information which is used to update the register. The main update is carried out annually as part of the system review, and particular pieces of legislation will always feature in the company's environmental objectives.

Scope of the management system

How far the management system should follow the life cycle of the product was also a point that was considered in depth for some time. Eventually it was decided that our system would initially be broadened to include raw materials when stored on site, the production and support processes on site, and the disposal of waste materials. This disposal of waste was considered to be one of two significant indirect aspects within the environmental management system, the other indirect aspect being the sub-contractors working on our site whom we expected to comply with our environmental procedures.

The decision of broadening the scope to include only two indirect aspects, and raw materials only when stored on site, was made on the following grounds:

- From a viewpoint during 1993/4 when everyone was at an early stage in the development of environmental control through implementing a management system, we found very quickly that not all stakeholders in our business held the same views or indeed had the same level of interest in environmental issues.

- In the early stages of development of our management system, we did not feel able to force our concerns or issues on other companies when not perfect ourselves – 'those living in glass houses etc.'

- Most raw materials or consumables were supplied by more than one company. Formulating partnership deals with one supplier for a particular raw material gave us a larger stake in each other's business. Business meetings arising from these 'partnerships' were found to be a good place to introduce mutual environmental issues for discussion without pressure or threat.

- Our customer base varies from large multi-nationals to one-man businesses, all of which have to adapt and work to different pressures and concerns. Making an Annual Report available to each of these businesses detailing our position on environmental issues, our objectives and how we monitor data, allows those with interest the opportunity for discussion .

- The concept of life cycle analysis is not yet fully developed into ISO Standards. For the sake of continuity with ISO 14001, this aspect of environmental management will be considered more fully when these become available.

Conclusion

Implementing the Environmental Management Standard proved a very useful exercise for several reasons.

Objectivity

Firstly, the methodology of the initial environmental review gave an 'objective' picture of what environmental aspects were on site, hopefully without the bias that could be expected from individuals working in isolation. Such individuals may be 'too close' or too involved in their own area of work and their complacency may lead to environmental aspects being missed. Alternatively, a subjective value could be placed upon an aspect's significance.

Thus we believed such objectivity would allow finite resources to be targeted so that real improvements would occur and benefits would be gained.

Environmental legislation

Having a process to recognize, and react pro-actively to, environmental legislation is one of the main benefits of the system. All too often, one visits lectures, seminars etc. where even after repeatedly being broadcast, the message still fails to get through. Five years after introduction of the COSHH regulations we were still coming across companies who were hearing of the regulations for the first time!

Impending changes (being pro-active)

Although decisions are made based upon present information, early knowledge of impending change allows a pro-active business plan to be drawn up which may well lead to improvement. In most cases, we have found that an 'environmental' reason for change will almost certainly be accompanied by a sound business reason. This may involve savings due to better control over processes, less waste etc. or it may mean the company does not fall foul of some (unknown or misunderstood) legislation.

A focus on targets

The environmental objectives developed from the Standard's Management Programme are valuable in ensuring that all other members of the company's management structure are focused on targets that are of value to the organization as a whole. If personal objectives and targets are already a feature of the company's management process, then additional environmental objectives will fit well within the existing system.

Finally, implementing ISO 14001 gives some indication to stakeholders in the business that environmental issues are being considered seriously and systematically. This may lead to benefits in terms of less direct regulation, reduced insurance premiums, fewer complaints etc.

Philips Lighting (Hamilton) Ltd

Written by Joyce Gauld, Environmental Officer, Philips Lighting Ltd, Hamilton, Scotland

Introduction

Philips Lighting Ltd is situated in Hamilton, Scotland, UK and manufactures lamps and luminaires. Several lamps are manufactured including tungsten lamps (normal household lamps, soft tone), sodium street lamps and fluorescent lamps. Production takes place in several buildings spread over a large site. The processes can be labour intensive and the main environmental impacts are glass and cardboard waste, special wastes, volatile organic compounds (VOCs) and noise. This note discusses the more difficult issues that were tackled during the Environmental Management System implementation.

Philips Lighting Ltd, Hamilton took a practical approach to installing a certifiable management system and let the documentation emerge from the actual situation on site rather than writing the documentation based upon hearsay or assumptions. Work started with an Initial Review being carried out to give us a snapshot of where we were starting from.

Gathering information for the review took 2-3 months as we wanted to monitor situations over a period of time to ensure that we had an accurate picture of what was going on around the site. The waste skips were the first to be monitored with daily visual checks to see exactly what was going into them. Skips were designated for different waste streams but this does not necessarily mean that everyone was using them correctly. The paperwork and documentation also needed collating and to be brought to a central point. This proved to be easier said than done! All the paperwork (transfer notes, consignment notes, etc.) was up to date but the people holding these papers in their files did not always know what they had so tracking it down took time. In our experience the disposal companies drew up the paperwork and just asked for a signature. This has changed and now we control the information on all outgoing paperwork as we know the composition of our waste better than anyone. The Initial Review was published with a set of photographs so we could not be accused of exaggerating. The length of time taken over the review was felt to be worthwhile. This was the starting point from which the Effects Evaluation Register was developed.

Our Effects Evaluation Register was set up using a spreadsheet. We split the processes up and by using a scoring system we were able to determine our significant environmental effects. This is not a static part of the system and will continue to evolve with the system. Some aspects of the standard were easier to apply than others and it is hoped our experience in some of these areas can be used to help other companies hoping to achieve certification. We achieved certification to both BS 7750 and ISO 14001 (although at this point in time, ISO 14001 was still in draft form – see DIS/ISO 14001 in Appendix I).

The legislation register

The Standard (BS 7750) outlines a requirement for a legislation register that is applicable to the site and for the information to be fed down into the areas where the legislation applies. What the site was looking for was a method of satisfying both requirements without presenting too much information for the departments to digest.

Although we required a register that gave us information on the legislation relevant to the site, we also wanted the document to be user-friendly. We wanted a document that could be understood by both managers and shop floor operators so that everyone who needed to comply with legislation was fully aware of their obligations.

Firstly, it was decided that the actual Legislation Manual should not just be a meaningless list of Acts, regulations, EC Directives and corporate policies. it was decided at this stage to identify all the relevant legislation and summarize the important points in the manual. This list was drawn up by going through all the available environmental reference publications. Many of the publications referred to the law as it stands in England and Wales and did not take into account the fact that Scotland has applied some of the legislation in a different way. Time had to be taken to check out the legislation and make notes as to where it differed for Scotland. Another briefing session agreed that in certain cases legislation applicable only in England and Wales *should* be included in the register so the reader could at first glance recognize that it was not relevant in Scotland. As the company has factories world-wide and a certain amount of information comes from a central point in the UK it was felt this addition was needed. These pieces of legislation were marked by an asterisk to show they did not apply directly to the site.

Another problem was that some pieces of legislation quoted in articles also sounded as if they should apply to the site but when checked showed that they did not. Questions were being asked about these so it was decided to summarize them as well and include them marked 'for reference only'. People could then see at a glance that we knew about legislation and the only reason for it not being deployed was that it did not apply to the site. Our own company memos often referred to legislation that again was not applicable in Scotland so by having a reference to it in the manual queries could be answered more quickly.

The next problem we faced was finding a way of passing down the information to the departments in a form that would be used. The summary manual was fine as a library copy but a copy for each production building was not seen as the best way of passing the information outwards to the rest of the site.

After discussions on what information each building required, it was decided to produce simple information sheets that were named 'Site Issues'. These sheets deal with one issue: for example, Disposal of Special Waste. The relevant legislation is listed (with a cross reference to the legislation register), along with the main duties/prohibitions required to ensure compliance with the legislation and how the factory meets its obligations (use of procedures etc.) As each building had its own 'issues' a smaller manual could be tailor-made for each building. With only 'relevant' information it was felt these booklets would be more readable and better used, especially by busy managers.

The method chosen proved to be successful. The production areas were grateful that time and thought had been given to providing them with a document that was easy to read while still ensuring they know what they had to do in order to comply with the law. If further information was required they could then follow it up by checking out the legislation register. The environment office holds copies of all major environmental legislation which could be borrowed if further information was required. If further information was required and the legislation was not available through the environment office it could be bought in or, if specific information was required, we could source it for them.

As a result, updating the information at departmental level was also easier. Any changes were easier to follow as the departments didn't have to wade through a pile of information – all they received was the facts they need.

Indirect effects/aspects

Our main problem was defining the indirect effects/aspects that related to the site and explaining to the managers why these were an important aspect of the system. It was generally felt that we had enough effects/aspects of our own to deal with, without taking on other companies' problems as well. Were our suppliers' environmental aspects our responsibility anyway?

We were not initially convinced that our resources should be occupied in considering the indirect environmental effects of our company. However, the benefits of considering indirect effects were gradually fully understood and thus we did indeed put some effort into this area. We considered the wider environmental effects: power stations for instance. They cause atmospheric pollution but industry in general needs to use electricity. Thus individual companies can have an indirect environmental effect by using less electricity through energy-saving measures. This saving has the added benefit of reducing costs – money which could be spent on other environmental improvements.

Effects of transport was another aspect to consider. In our evaluation system each part of the process is given a score based on certain criteria (quantity of raw materials used, toxicity, waste, nuisance, etc.) These scores are added to give a total for that part of the process. This total is then multiplied if there is legislation pertaining to it to give a final score. In our system if this total is over 100 it is deemed as significant. The way in which our Effects Evaluation Register (EER) was set up meant that individually (per item delivered to or removed from the site) the scores were not that high but once we added all the transport scores it was showing as a significant effect. Goods come into the factory from all over the UK and Europe so any project set up to deal with this issue will be a major one spanning years as opposed to months.

The supply chain and its effects on our site was another major problem for us. How far back in the supply chain do you go? Do you have to consider all potentially environmentally harmful issues in the supply chain? Is the problem 'ours'? How much influence could we have anyway? All these questions needed to be addressed before we decided how we would tackle where the supply chain started and ended.

Gaining management support, and commitment, especially in the purchasing function for using our resource of time to investigate the supply chain had its difficulties. The support was more forthcoming when it was pointed out that if part of the supply chain was environmentally sensitive and the process was not being managed properly, it may be closed down by the authorities. Then *we* would have a problem as the supply of material to us would cease.

In the end it was decided to keep an open mind and go as far down the supply chain as we felt was warranted by looking at:

- How much material we use

- How much influence we could have

- Whether we could source from a different supplier

The process was started by sending our immediate suppliers a questionnaire which also asked about their suppliers. These were subdivided into groups of like suppliers (for example chemical, plastics, metals). This gave an overview of the industry type as well so we could immediately see if one company had not filled in the form correctly – these were the first to be followed up by phone calls and visits.

Co-ordination between the quality, purchasing and environmental departments was required so that our customers were aware of our combined requirements. Visits to customers were therefore more effective as all combined specifications were discussed.

Raising awareness and training

On a site with approximately 900 staff working different shift patterns and 24-hours-a-day, 7-days-a-week working, it was not possible to get everyone together at one time to give them awareness training. Our awareness training was developed so as to reach everyone on site and was carried out over a full week. Personnel were targeted at either the beginning or the end of their shift and this meant, for us, working nights and over the weekend to make sure everyone had seen and understood the material.

We had the help of two environment students from the local college and they were very willing to get involved in the project. After some discussion it was decided that the content of the awareness training would be split between general environmental issues and what the factory was doing with regards to BS 7750/ISO 14001. General information was gathered from environmental groups (both local and national). Some groups even gave us posters to use. Other leaflets were produced on site and gave information more relevant to the site (for example, why we could not put chemicals down the surface drains). We also produced a fact sheet on BS 7750/ISO 14001 and printed copies of the environmental policy statement.

The next question was: 'In what form do we give this information to personnel?' After meetings with the building managers it was agreed to set up an awareness 'table' which would be located for a set time in each of the buildings. The staff would be made aware of the times it would be in their building and when they could go and look round. The students agreed to help man the table for us so that personnel could ask questions if they wanted. It was felt this was better than a lecture style talk as personnel felt more at ease asking questions on a one-to-one basis. The objective of this was to increase knowledge and not to make personnel feel uncomfortable if they did not know or understand about a certain issue. In a large crowd people are less likely to ask questions. This proved to be a worthwhile step and many questions ended in lengthy discussions. The students' background meant that they were more than able to answer the majority of questions put to them. However, if anything did confuse them then the environment officer was called in to help.

We found this method of raising awareness to be very effective. Personnel who had outside interests such as fishing were interested in water pollution. This started the discussion and then led on to what the factory was doing in relation to this. The method gave a better understanding as to why they had to apply a procedure, not just stating it was better for the environment. They also liked the fact they could take leaflets away to read but were not being pressurized into doing this. They were absorbing as much information as they wanted and were also feeling as if they were part of the improvements on site.

Specific training for individuals was straightforward as we already had a well-structured training department. Key staff working in significant areas were targeted first along with senior managers and production managers. Initially three days of training were set up in-house with the environment advisor from our corporate headquarters in Eindhoven and the site environment officer. The first day targeted the senior managers, the second production managers and other key staff and the third process operators. The third day's training was tailored to be more process specific and all three days went well.

Training programmes were set up to train environmental auditors. These training sessions were carried out by an outside consultant and followed by a one day in-house course covering:

- The site's environmental system

- The workings of the effects register

- Our local policy

- A site tour of the areas significant in environmental terms

The training programme is ongoing and the environment is now a topic covered in the existing training programme of our tungsten lamp making section. Other departments have requested specific training and this has been carried out by the site environment officer and the personnel in each area responsible for environmental issues.

Training will always be ongoing but in time it is hoped that every member of staff will have received more in-depth training on environmental issues as a follow up to the awareness training, although this will obviously depend on the resources available.

Chemical waste handling

The most difficult challenge in this area was to get personnel to change their attitudes and working practices with regards to waste chemicals. On a large site with different business groups the waste chemicals are collected at a central point. It is the environmental officer's responsibility for correctly disposing of this waste. The individual work areas see this material as waste and therefore it doesn't matter how it is contained as long as it reaches the storage area; labelling wasn't thought to be an issue as it was frequently marked 'waste'. This wasn't too helpful when a correct description was required for the consignment note! Meetings were set up with the departments involved and it was explained to them why the chemical wastes could only go into certain containers and that the labelling had to be correct and there had to be *one* label per container. Both the format and the information required on the labels were agreed between the departments and the environment officer so that they were acceptable and, more importantly, understood by everyone.

Legislation has changed so rapidly over the last five years that, understandably, it is difficult for personnel to keep up with it – especially if the subject matter is unfamiliar. This was where the 'Site Issues' sheets became very useful. The personnel dealing with the chemicals had a simple document that they could understand and that explained why we had to change our working practices.

Changes do not happen overnight and personnel do not quickly change the way they have been working for the past 20 or more years. Often things will have to be redone or discussed so a better method of implementation can be found. It can be left to a few personnel to check and recheck the situation to ensure the changes become the preferred method of working. Training is vital and help (for example, checklists for the departments) are important to show personnel that they do have backup and are not being left to take on 'another job'. Our experience has shown that people work better and are more receptive to new ideas if they feel part of what is happening and have backup from someone if it is required. Any checklists that were made up were done so with the involvement of the personnel who were working in that area or process.

When to get in touch with the certification body

'At what stage of the implementation process do we begin to involve our chosen Certification Body?' was a question that we asked ourselves. Options varied from when the manuals started to evolve (or earlier) to when the system had been in place for several months. We finally decided on the time the manuals were being written up. By then, we had in place our initial review (although not part of the certification process), effects evaluation register, our significant effects register and the legislation register. Our improvement plans for the year had been published and work had started on them. Records had been collated and archive material had been gathered and filed. By the time the certifiers came in our records had been running for 18 months.

The starting point in the certification process was a Pre-Audit which proved to be very useful to us. A lot of discussion took place and we later actioned some of the points raised. As our manuals were not complete at this stage we felt we would rather have another one-day assessment once they were complete before going forward to the certification audit. Again, we gained a lot from this visit and it also made us confident that we were on the right track.

We then went forward for the certification audit a few months later and were successful in gaining certification to both BS 7750 and DIS/ISO 14001.

Although we, as a company, had a corporate requirement to achieve certification, we have none the less seen it as a major achievement for the factory and one that has included everyone in the workforce. The certificates are only the starting point for us (a bit like learning to drive once you have passed your test) and we hope to see many improvements over the next few years that will improve our environment.

Toshiba Consumer Products (UK) Ltd

Written by members of the Environmental Team, Mike Cockerill (Senior Manager, Manufacturing Engineering) and Nick Taylor (Industrial Engineer), Toshiba Consumer Products (UK) Ltd, Air Conditioner Division, Plymouth, Devon

Introduction

Toshiba is situated near Plymouth in the South West of the UK, in a rural area and adjacent to a National Park. Toshiba was established in Plymouth in 1981, with the purpose of assembling microwave ovens. It moved to its present site in 1984.

During 1990 and 1991, the site was further developed to enable a change of use to take place (this was the manufacture of air conditioning units for commercial use). The redevelopment was carried out with additional facilities to satisfy the requirements of the Environmental Protection Act of 1990 for the processes and plant.

The new product range required the storage and control of HCFC refrigerant gas R22 (a less ozone-depleting gas than CFC gases). Additional plant included an aqueous pre-treatment plant, a solvent-based paint facility, a steel press shop and an aluminium and copper heat exchanger fabrication unit. A solvent stripping plant was also constructed and equipped with an activated charcoal vapour recovery system to control atmospheric and effluent emissions.

During the phase of site development, strong links were established with the local authorities. This ensured that later authorizations required for plant operation and consents to discharge from the effluent treatment plant were granted. The factory unit itself was built and previously occupied by an engineering company engaged in machining, heat treatment and chrome plating of metals. During the site development we found evidence of ground contamination from this previous use. This was fully surveyed and assessed for environmental impact.

The process of air conditioner manufacturing involves many stages – some labour intensive, others automatic or semi-automatic. Raw materials used include copper, aluminium and sheet steel and processes include metal stamping and pressing, welding and metal working, cleansing in solvents, and final painting, as well as incorporation of refrigerant gases into the product.

Establishment of an Environmental Management System

In April 1995 Toshiba Consumer Products – Air Conditioner Division responded to a Toshiba corporate plan for all Toshiba sites, globally, to be certified to an environmental management system. BS 7750 was specified as the Standard to be certified against, to be replaced by ISO 14001 when it was published.

The Plymouth site was chosen as a pilot exercise for other Toshiba operations in Europe due to the environmental significance of its processes. The time-scale demanded by corporate headquarters in Japan was ambitious, the requirement being to obtain certification within twelve months for the Plymouth site. There was, however, a strong belief that due to the existing high levels of operational controls and environmental awareness amongst the majority of plant personnel, this time-scale was achievable.

Although Toshiba Consumer Products (UK) Ltd. is a division of a large multinational organization, the specialist expertise required for implementing such a system was not readily available within the organization and so personnel had to undergo a steep learning curve. Thus, in many respects, the plant faced the same difficulties faced by many smaller organizations. Resources, in the form of management time were also stretched due to the everyday pressures on time placed on all staff in running a busy, dynamic business.

This case study highlights the main challenges facing Toshiba in the implementation of its environmental management system which was certified to BS 7750 in April 1996, and later 'converted' to ISO 14001 during a routine surveillance visit by the certification body.

The most challenging areas of the Standard are described below.

Challenge 1: Environmental aspects identification and evaluation

This was certainly the most difficult task: to define environmental aspects and evaluate their significance. Brainstorming sessions were held by an 'environmental team' to answer such questions as:

- What indeed is an environmental aspect?

- What is a direct versus an indirect aspect?

- How can one aspect be more significant than another?

Although Toshiba had a Quality Assurance Management System (certified to ISO 9002 in 1994) and therefore had experience of working to structured management systems, this experience was of no assistance in answering the above questions. (There is no equivalent clause within ISO 9000 systems to address environmental issues.)

Identification of environmental aspects
Through the brainstorming sessions, hundreds of environmental aspects were eventually identified, as the manufacture of air conditioning units is complex, involving many raw materials, sub-assemblies and processes. From this long list, the environmental team identified many areas that had some environmental significance attached.

Evaluation of environmental aspects
It was therefore clear from the outset that a simple to use, easily understood evaluation system, with a clearly defined system of weighting impacts, was required. This system had to be able to differentiate the significance between totally different environmental issues. For example:

i) Cleaning with trichloroethylene

ii) Purchase of copper pipes

iii) Split system air conditioning systems

The areas that needed evaluating were initially broken down into three broad groups of perceived environmental risk:

1   Materials

2   Processes

3   Products

Criteria were then sought to identify the severity and potential environmental severity for each aspect with typical parameters being:

*   Resources required

*   Public pressures

*   Location

*   Costs

These questions, and others, were then matched to the three broad environmental risk groups (as above). A numerical system of rating environmental significance was decided upon, the answers to the questions being given a number, ranging from:

    5   For a significant impact

to:

    1   For an insignificant impact

Key areas were investigated to identify significance. Typical areas included:

- All COSHH (Control of Substances Hazardous to Health) records and material safety data sheets

- Process authorizations made with the local authority

- Consents to discharge effluent, made with the local water company

- Purchased materials

- Manufacturing processes

- Products sold to consumers

- Bought-in products

- Wastes disposed of to landfill

To ensure a common approach to evaluation, and to reduce any possible bias due to familiarity with the processes, one engineer was responsible for all the initial evaluations. He would be aware of most of the processes but not the detailed environmental aspects of those processes. However, in some complex areas, input and guidance was sought from other staff. These initial evaluations were then scrutinized by peers for a measure of 'common sense'. A final judgement was given by the management representative through a formal review process.

It was found that if, for example, the evaluation of a raw material resulted in a high environmental risk score, then the amount of the raw material used in the process also had to be evaluated for environmental significance. Thus a highly hazardous raw material used in small amounts can be just as environmentally significant as a material used in high quantities but with a low hazardous nature.

The amount of time involved in evaluating the above risks was reduced by the process of grouping of products. 90% of the product range were merely larger or smaller variants of one standard model. Thus this smaller number of groups were evaluated for environmental significance.

The environmental team thought that the process of identification and evaluation of environmental aspects had been performed without any bias or pre-judging of significance. However, during the pre-audit, the certification body auditors queried whether the team had indeed identified every relevant environmental aspect prior to judgement of significance.

For example, the team had decided that the use of paper clips in the offices were of such a trivial environmental impact, that they were not entered into the aspects register. The team was then placed in the uncomfortable position of justifying this to the external auditors. The external auditors, although agreeing with our decision that paper clips were an insignificant aspect, posed a question. This was: Could the Team's subjectivity also discard other environmental aspects that may have significant implications? Perhaps, if not now, then in the future. The auditors had to be sure that the environmental management system was capable of picking up all environmental aspects without bias as to their significance.

Now, all environmental aspects are evaluated on a regular basis, as due to changes in legislation, suppliers, packaging, production volumes etc., the environmental significance of an aspect could radically alter. The environmental team now believes that the system is strong enough to identify all aspects and that none has been overlooked.

Challenge 2: The concept of 'abnormal' operation

In the guidance material within the annex to ISO 14001, under section A.3.1 'Environmental Aspects', there is a clear indication that the environmental aspects evaluation should consider normal operating conditions, shut-down and start-up conditions as well as the realistic potential significant impacts associated with reasonably foreseeable or emergency situations. (Unfortunately, the guidance material does not expand upon what is meant by 'reasonably foreseeable'. We have interpreted this as being the exercise of due diligence by having in place inspection and maintenance procedures.) There is also a reference that during a preliminary environmental review consideration should be given to abnormal operations within the organization.

Normal and emergency situations were well understood by most personnel, but what constituted an 'abnormal' or 'foreseeable' situation was not. This was a large area of uncertainty which required much thought. Eventually, the thoughts of the team became more focused and the following example demonstrates the approach taken.

Introduction
Part of the painting process for the air conditioning units is the water-based cleaning and treatment of steel and zinc plated components using an alkaline degreaser/cleaning agent. Once cleaned, the components are treated with a solution of zinc and nickel in a phosphoric acid solution.

The problem
Under 'normal' operating conditions, the waste solutions from the above process are discharged to sewer after suitable treatment in the effluent plant to meet the discharge consents agreed with the local water company.

Operational controls are in place to ensure that such discharges meet the consent and therefore under normal conditions, the environmental aspects are insignificant.

An 'emergency' scenario was relatively easy to visualize. The effluent plant operates by removing all oil and solids, and monitors and controls the pH of the effluent at several critical points. For example, if the monitoring controls failed, allowing the dosing of neutralizing agent to be insufficient, the pH would drift out of specification and several alarms would be triggered. In the meantime, the outflow of effluent exceeding the discharge consent would continue at 3000 litres per hour (the operational flow) until the process was stopped by the operators. Thus in this situation, the environmental impact would take on an extremely high significance (that is, exceeding the discharge consent).

To complicate the situation, the designed throughput of the plant was set at 1000 litres per hour, which meant that the technology was being used at the limits of its design.

The costs associated with responding to such emergencies were high – the more intricate and fool-proof they became, the more money and space they required for installation. The problem was always the flow at 3000 litres per hour.

The solution

The abnormal operation was therefore defined as the period of operation when the dosing of the neutralizing agent went wrong, for whatever reason, but the outflow of untreated flow was kept within the boundaries of Toshiba effluent plant (by the procedure of operators switching off three pumps). This would allow an abnormal operation of several hours before an emergency situation was declared. The abnormal operational control procedure also called for a waste tanker to be sent to the site, by our waste disposal company, to pump out the effluent. The procedure ensured that such a contract with the company was drawn up to respond to such abnormal situations in a short space of time.

The environmental significance for abnormal operation is now rated as medium: that is, above normal operation but not as high as emergency operation. Thus, the same environmental aspect has three different levels of significance depending upon the situations of normal, abnormal and emergency.

Other staff in the paint shop – maintenance staff and support engineers – are also familiar with this procedure and have their part to play. Because an abnormal situation is just that – that is, it happens infrequently – responsible personnel may not remember the correct procedure to follow: who to contact; what actions to take, etc. Therefore, such procedures are affixed in prominent places to remind staff so that the procedures are followed effectively when the time comes.

The procedures have been tested as far as practically possible, and have worked well in such, admittedly artificial, scenarios.

Conclusion

The team believed that the identification of environmental aspects had been performed with a sufficiently wide scope and that evaluation had taken place without bias using as objective a methodology as possible. This methodology was shown however, during the external audit, to be just slightly flawed. Nevertheless, the systematic approach undertaken was fundamentally sound and, therefore, most of the time, the team could offer logical reasoning in reply to questions from the auditors.

The team learnt that once the initial identification and consequent evaluation of environmental aspects has taken place, the system must be dynamic to ensure that there is ongoing control and reduction of such environmental impacts. For example, since the process of environmental improvements began, the team have achieved the objective of phasing out the use of 1,1,1 Trichloroethane for cleaning and degreasing purposes and replaced it with trichloroethylene (in accordance with the Montreal Protocol for ozone-depleting substances). A further building on an adjoining site has been purchased (for expansion of production). In both cases, an environmental aspects identification was performed. That is, the change to trichloroethylene brought a different set of aspects and the additional building raised the issue of contaminated land (due to previous use).

Operating the process under abnormal conditions, and recognizing a different set of aspects in such conditions, is now an area where knowledge has been gained and is being further developed within the environmental management system. Further, the evaluation system that was devised has now been developed in scope to include life cycle analysis of all the products.

Chapter 7

The auditor

Introduction

ISO 14001 was designed to be an auditable standard and, within this book, references have been made to 'certification bodies' and 'third party' auditors who will undertake this auditing process. Having committed themselves to obtaining ISO 14001 certification, organizations naturally enough focus on gathering environmental information, allocating resources and employing environmental consultants (if required). Budgets will have been prepared, and personnel coached in how to answer the auditor's questions. A tremendous amount of preparation will have been done to enable certification to be obtained within the planned time-scale. However, it could be that no face-to-face contact has been made with the representatives of the certification body (that is, the auditors) until implementation is well in progress and the deadline for the pre-audit is fast approaching.

Although some information is available from certification bodies setting out the certification process (see Chapter 4), little or no information is readily available about the actual auditor, or team of auditors, who will visit the organization's site to perform the audit. This individual, or team, will need access to all departments, will disrupt normal organizational activities, demand valuable management time, interview (possibly very nervous) staff, and will have the power to grant certification, or otherwise to the organization. So it is wise for an organization to do just a bit more preparation and to find out what sort of individual (or individuals) will actually arrive on the first day of the pre-audit. Such preparation can only be a good investment of time so that a good rapport can be established with the auditors as quickly as possible. A good relationship will be essential between client and auditors so that:

- Disruption to the business is minimal

- No differences of opinion or interpretation of the Standard leads to animosity

- Personnel do not feel threatened or intimidated

- The audit proceeds smoothly

The auditor, or auditors, will have to perform a physical 'walkabout' of the site, ask questions, take notes, and make as objective a judgement as possible on the sample of the environmental management system that is presented in evidence. Advances in video techniques and interactive information technology may one day change this but, for the foreseeable future, this appears to be the only (if somewhat imperfect) method of verification and relies heavily upon the concept of a 'standardized auditor'.

A standardized auditor cannot exist, because being human, the auditor like everybody else has a unique set of characteristics (including fallibility). However, some of the processes used by certification bodies to 'calibrate' auditors will be described. This calibration ensures that a minimum level of consistency of auditing is achieved.

Auditor characteristics

Organizations implementing ISO 14001 may have been certified to ISO 9000 for some years and will be familiar with being visited by a quality assurance auditor on scheduled surveillance visits. Unfortunately, because of quality assurance's long history, it has received some 'bad press' as well as 'good press'. Some myths have grown up, and been perpetuated – some unsubstantiated; others, unfortunately, having some substance.

As in any myth, stereotypes and caricatures emerge. For example, all quality managers will remember the laid-back, affable auditor who skims the surface of the quality system, is readily distracted by endless offers of coffee, is relieved when a trivial nonconformance is discovered, and departs satisfied that the objective of auditing has been achieved (that is, raising a corrective action request).

The extreme opposite is also well known. This is the arrogant auditor who creates havoc by the single-minded pursuit of an area of weakness (probably trivial in nature), enjoys verbal confrontation with the quality manager, and departs with a freshly reinforced ego – leaving staff in the organization thoroughly and totally demoralized by such a negative style of auditing.

Perhaps it is fair comment to say that auditors, as an essential characteristic, must possess a sense of humour to endure such tales that abound, of their chosen profession.

In reality, most auditor behaviour is somewhere in between the two extremes. Auditors have to be reasonably confident in what they are doing and should present themselves to the client as assertive. Unfortunately, in the stressful situation of an audit, this can be misinterpreted as arrogance. Almost every other day, auditors are facing new clients – complete strangers to them – and they want to get the best out of the person being interviewed as quickly as possible. Upsetting people will create antagonism towards the auditor. The willingness to answer the questions honestly and in an objective and helpful manner will be lost. Personnel are often quite defensive when being interviewed about their tasks so putting them at ease is an essential requirement. An auditor, therefore, needs to be able to relate to all sorts of individuals and also be a good listener. Auditors need excellent interpersonal skills and need to establish a good working relationship with the client very quickly. Time is usually against them – due to commercial considerations – and so, during the audit, 'small talk' may be at a minimum.

Auditors, like all employed personnel, are selected via an interviewing process in which interpersonal skills are assessed as well as the more tangible qualities of training, education and experience. Auditors, with these skills and personal qualities, are looking for the level of compliance with the Standard, and not the extent of noncompliance. Auditors also want to add value to the audit. To focus on purely negative issues and raise corrective action requests does not give job satisfaction. The auditor would like to feel that value has been added to the client's system, through the mechanism of observations to either improve or simplify the system, and yet still meet the requirements of the Standard.

The client, rightly or wrongly, expects the auditor to know their (the client's) particular business immediately. In an ideal world, the auditor could spend several weeks at the client's premises, learning the processes, getting to know personnel and their responsibilities as well as assimilating the unique language and culture that any business possesses. Unfortunately, no client is willing to pay for this so the auditor must be able to assimilate this information very quickly, on-site, as well as performing the audit effectively.

Finally, of course, auditors do need regular cups of coffee, the same as everybody else.

Auditor qualifications

As previously suggested, the Certification audit may require an auditing team to ensure that the appropriate skills and experience to perform the audit are present.

The team approach needs some further explanation. It is not meant to be a show of force to the client, although it may appear that way to some. It is a fact that, at these early stages of environmental management systems development, the required expertise to conduct a fair and valid assessment will probably not rest in one individual. The individual who has expertise of all national and international legislation, and has had practical industry experience in all industrial sectors, and is also an experienced environmental auditor does not exist (ISO 14001 was only published in final form in September 1996). The years of experience and training, the tremendous amount of information to be assimilated, as well as the requirement to be continuously up to date on the many related environmental issues, would place unrealistic demands on any one individual. Therefore the team will comprise of individuals who, collectively, will bring the correct expertise to the assessment. They will be led by one of the team – designated the team leader.

However, it is understood by all interested parties, notably the certification and accreditation bodies, that such individuals will broaden their expertise (that is, acquire new skills via briefings from experts, specific training and participating in audits with technical experts). For example, an environmental management systems auditor, with a scientific education, training and industry experience, will readily assimilate knowledge of forestry practices by private study and, of more importance, by being on pre-audits and certification audits with forestry experts. Such an auditor will be able to put such specific forestry practices into the context of the (generic) Standard, and perform a valid audit of the forestry organization's environmental management system. This does not of course create a new forestry expert but it creates an environmental management systems auditor who, in the future, will be able to add value to forestry management environmental systems generally, by virtue of his experience.

The qualifications and skills required within the auditing team are four-fold:

i)   Management system auditing capability

ii) Environmental competence – including regulatory and legal compliance

iii) In-depth knowledge of the Standard – ISO 14001

iv) Technical knowledge of the industry

These qualifications are described below.

Management system auditing capability

Accreditation criteria demand that certification can only be granted on the basis that the environmental management system is capable of delivering regulatory compliance and environmental performance improvement.

In other words, it is the strength of the management system which is being considered and those skills which can examine the effectiveness of the system in achieving this are paramount.

It is a fair comment to say that most existing environmental system auditors will have gained auditing experience through auditing of ISO 9000 quality assurance systems. This will have given them experience of knowing how to phrase questions to get the best answers from personnel being interviewed. They will be familiar with the structure, composition and hierarchy of documentation of a management system and, as discussed in Chapter 5, there are many commonalities in the philosophy and structure between ISO 9000 and ISO 14001, for example.

It is also likely that they will have a wide experience of auditing to ISO 9000 organizations with activities relevant to environmental control. Such activities could include the operating of:

- Landfill sites

- Waste disposal (collection)

- Municipal waste incinerators

- Water treatment plants

- Effluent plants

This will provide a good foundation for ISO 14001 assessment skills. Furthermore, in the UK, some auditors will also have auditing experience of BS 7750 (both 1992 and 1994 editions).

Environmental competence

Environmental competence means the possession of sufficient experience and knowledge to identify readily an environmental aspect and to judge its level of significance.

It is also clear that a good knowledge of pertinent environmental issues and legislation is necessary so that the auditor can judge whether the system being audited will deliver performance improvement as well as regulatory compliance. The need to exercise such environmental judgement, or competence, is proportional to the environmental profile of the situation. High-level competence is often required in the chemicals or power generating sector, for example, where the potential for environmental incidents are higher, environmental issues are broader and far reaching and tighter legislative controls are in place.

The auditor needs to be able to identify an environmental aspect when one is presented and to judge its order of significance. Being aware of, and understanding the current techniques for control and mitigation of such an aspect is also a measure of auditor competence. When judging the order of significance, therefore, the auditor will consider all the aspects of:

- Atmospheric emissions

- Discharges to aquatic environment

- Waste management and disposal

- Contamination of land

- Effects on eco-systems

- Nuisance pollution

The auditor must also have a clear understanding of the indirect aspects of the organization being audited. Indirect environmental aspects are the aspects particular customers and suppliers of the organization being audited are likely to have.

Auditors themselves will possess relevant professional and technical qualifications such as botany, biochemistry, chemistry, forestry management etc. and will have had structured training in environmental issues. Although minimum level qualifications are suggested in ISO 14012 (see Appendix III) and in Accreditation Criteria guidelines (see Appendix IV), most environmental auditors will possess a degree-level qualification and, in many instances, a post graduate degree in an environmental discipline.

In-depth knowledge of the Standard

Obviously, the auditor needs to know the Standard intimately – this is a fundamental requirement for effective auditing. Such an in-depth knowledge can only be gained by studying the Standard, including the Annex, and applying that knowledge in practical auditing situations. Reading of the Standard can of course be done in isolation but auditors will have also participated in workshops with colleagues to discuss areas of interpretation. Interpretation in this sense refers to words, phrases, sub-clauses or clauses within the Standard that suggest different meanings to different auditors. Practical experience is gained whilst auditing under supervision for a set number of audits (the minimum number of supervised audits is defined in the Accreditation Criteria).

Technical knowledge of the industry

The 'language' of the industrial sector being audited needs to be understood by the auditor. Knowledge of current best practice, with regard to environmental control within a particular sector, helps to ensure a meaningful audit. The need for such specific technical knowledge is also proportional to that sector's environmental profile. For example, the water industry would require a more detailed technical knowledge than the packaging industry. Such knowledge will have been gained through previous operating experience within that industry. Most environmental auditors will have had previous employment experience though not necessarily in an audit-related function.

In order to satisfy accreditation criteria, a member of the auditing team who can offer such industry experience must be present at the certification audit. Clearly, this industry experience must be reasonably substantive. What is required is some level of expertise gained by operating at a technical or managerial level for a reasonable period of time so as to assimilate the 'norms' and working practices of that business sector.

The technical expertise requirement for a team composition is carefully controlled by the certification bodies. This level of expertise requirement is carefully monitored by the national accredited body. Auditors are required to provide documentary evidence – which may be scrutinized by the accreditation body – as to why they as an individual, and as a team, are capable of performing a fair and valid assessment for each particular client. Such documented evidence addresses the previous requirements (i) to (iv); it also acts as a safeguard for the organization being audited, ensuring that they are getting auditors who as a team will be aware of the norms applicable to their industry. This evidence also ensures that eventual certification is meaningful and worth having.

To sum up this section, third-party auditors require more than a modest level of education, training and experience in order to perform their jobs effectively. Clients, themselves, should enquire as to the environmental background of the auditors. Although auditors should be pro-active in this respect and tell the client something about themselves prior to the audit, the opening meeting is an ideal opportunity for such enquiries. There are voluntary registration schemes for environmental auditors (see Appendix II) setting training, experience and qualification criteria. However, it does tend to be the certification bodies who set minimum requirements – as they have to be mindful of the fact that they are audited by accreditation bodies and could be taken to task if their auditors do not demonstrate a certain minimum level of competence.

The 'hybrid' auditor

There is a school of thought which states that environmental management systems auditing is a hybrid discipline in that it requires an individual auditor who can function both as:

1   An **environmental auditor** who is experienced in auditing a client's environmental performance including environmental impact assessment of new ventures, due diligence auditing, legal compliance and being an expert witness in legal disputes. As such, this auditor will be technically educated in an environmental discipline, have wide experience of environmental issues and could also be classified as a technical expert.

2   An **environmental management systems auditor** who is experienced in auditing documented environmental management systems against written standards such as ISO 14001. Some of this experience may also have been gained from auditing quality assurance systems. This type of auditor does not possess the depth and breadth of knowledge of the technical expert above but is very able to verify whether the mechanisms, the controls and the managerial activities of an organization are robust enough to deliver environmental performance improvement. Of the two types of auditor, this one is prevalent within certification bodies.

It is important to recognize the differences between these two. Some examples of these differences could be:

A pollution incident
The environmental auditor will be well-versed in the wider environmental issues and may see a pollution incident purely as poor management of the process.

The environmental systems auditor will be looking to see whether the organization has recognized its weakness on this occasion, and has strengthened its management systems to prevent this occurring again. Any noncompliance would be against the lack of corrective and preventive actions, rather than the pollution incident itself.

A legal noncompliance

An environmental auditor could regard a legal noncompliance as a non-compliance in itself. The truth is that no company is 100% compliant with legislation 100% of the time.

An environmental management systems auditor would look for the causes of failure of the system which prevented the delivery of regulatory compliance. Assurance would be sought, through the mechanisms of corrective actions, and objectives and targets, that the situation would improve over a period of time, so that such failures become less likely.

Evidence of abatement technology

An environmental auditor may seek excellence in abatement technology, and look for the application of BATNEEC and EVABAT (see Appendix I).

The environmental systems auditor however will seek a management program to improve current abatement technology.

It follows that after a period of time the two fairly distinct types of auditor will merge as both disciplines rub off on each other and expertise is assimilated. This, of course, places much more responsibility on the shoulders of the one hybrid auditor as the conflicts noted above have to be resolved within the one individual. At the present time, of course, the team leader (who will be an environmental systems auditor) has to balance the findings of the technical expert with the management system findings and reach a verdict on compliance with ISO 14001.

The evolution of the hybrid auditor will not negate the team approach but it may be that for all but the most complicated and involved processes, with resultant complicated environmental impacts, using only one hybrid auditor will emerge as the most appropriate approach. Going a step further, with the development of integrated management systems (as outlined in Chapter 5) that encompass quality assurance, occupational health and safety as well as environmental management, the evolution of the multi-disciplined systems auditor may well become a reality.

Auditor methodology

Going back to the theme developed in the section describing auditor characteristics, there has been, it must be said, some emphasis on the structure of system documentation for quality assurance (ISO 9000), from both consultants designing the system, and certification bodies auditing such systems. A fully documented system with minutely-detailed procedures and work instructions, all cross referenced, individually signed and dated by the chief executive, use of coloured paper (to stop unauthorized photocopying) and every member of the organization having a controlled set of manuals and procedures, was considered a 'norm', and a situation guaranteed to make auditors very comfortable. The ability of the system itself to deliver true quality and customer satisfaction, was quite often lost in the morass of documentation.

Certainly, some external auditors when faced with a mountain of documentation experience a certain 'feelgood' factor. Such an amount of documentation is, of course, very tangible and readily audited (such auditors may well struggle to 'feel good' with the trend towards less paper and more use of software-based management systems).

Presenting such a top-heavy documented system to an environmental management systems auditor will not elicit the same response. First and foremost, the methodology of auditing is different. The auditor needs to ensure that the system is based upon: the management of environment aspects; delivery of environmental performance; and improvement and delivery of regulatory compliance. The decision-making processes, how environmental aspects are given a measure of significance, and whether strong linkages are apparent are uppermost in the auditor's mind. He will consider:

- Aspects identification

- Policy

- Objectives and targets

- Responsibilities

- Programs/procedures

- Internal audit

- Review

Organizations may well focus too much on the environmental management system manuals and procedures as the main element of the total system. A balance is clearly needed between such a documented system and the improvement techniques and methodology. This is reflected in the Standard's Annex and *Accreditation Criteria Guidance*.

Likewise, it should be kept in mind that the environmental auditors are not looking for a perfect system. They are looking for a workable system – a system that is robust enough to withstand the many other conflicting priorities of a business trying to succeed in today's tough commercial world. Auditors are also looking for a system that is capable of giving performance improvement. During the pre-audit, the auditor has a certain amount of flexibility in reporting improvements, development points and advice to the client; thus the client understands clearly what is required for the next stage of the certification process and movement to the next phase of the certification process can occur (see Chapter 4). However, the auditor is not allowed to *design* any part of the management system. This is classed as consultancy and is forbidden by certification body rules and is rigorously monitored by both the certification body and the accreditation body. At the certification audit, the auditor has less flexibility and can only raise observations as a mechanism for system improvement. Having said that, the auditor, even if raising a corrective action request, will ensure that the client understands what is required to correct the problem. The solution, however, must come from the client and not the auditor.

Environmental auditing, the law and auditors

It was never the intention of the Standard for auditors to become regulatory authority spies. Certain countries do however make it the law for anyone becoming aware of a breach in regulations to inform the regulatory authority. This places the auditor in a dilemma. He must choose between:

• Compliance with the law of the land

and:

• Respect for the confidentiality agreement that is made with the client

If an organization is discovered by the auditor to be in breach of regulations, then the auditor will investigate further. Answers will be sought for the following:

- Has the breach been identified by the organization itself? (This would indicate the strength, or otherwise, of the internal audit.)

- Is the occurrence relatively isolated in frequency?

- Have preventive measure been evaluated by the organization?

- Has dialogue taken place between the organization and the regulatory authority?

If the answers to the above are all affirmative, then the auditor can be convinced that there is a commitment to compliance. The regulatory body will not be notified in this case and so the client's confidentiality is not compromised.

Normally, when an authorization to operate a process is granted by the regulatory authorities, the authorization states, as a condition, that breaches must be notified to the authority. Thus, the auditor should ensure that the organization's system records any infringements of regulations and that the appropriate corrective action is taken.

Consultants

A brief mention is made at this point of the consultants available to assist organizations to achieve ISO 14001. As previously indicated, certification body auditors are limited as to what advice they can give to clients in order to ensure no conflicts of interest arise during the certification audit. Therefore, those organizations without the correct resources available internally will need to call on the services of an environmental consultant. However, as with environmental auditors, there are two distinct types of consultant available:

- Environmental consultants

- Environmental management systems consultants

Each type of consultant is described below.

Environmental consultants

These will probably be an established company, partnership or even an individual, who have been in the business of environmental impact assessment for many years. They will have performed several different types of audit for different reasons (expert witnesses for example) and as such are well versed in all aspects of environmental legislation, land or buildings contamination, impacts on flora and fauna etc. It may well be that over more recent years they have extended their portfolio into performing preparatory environmental reviews for implementing organizations, especially since 1992 when BS 7750 was first published.

They may not have, as yet, the expertise in assisting companies through the management system element required by ISO 14001 but this is not where their strength lies. Their strength, and where they are best employed, is at the preparatory environmental review stage. If an implementing organization does not have the in-house resources to perform the identification of all environmental aspects required for the preparatory review, then it must seek outside help from such technical experts.

Environmental management systems consultants

Again, these may be an established firm, partnership or individual but they are very likely to be experienced quality assurance consultants who have spread their portfolio into environmental management systems consultancy. They will have many years experience in working with informal management systems and developing them into formalized, documented systems.

As the ISO 14001 Standard is so new, they will not have many years' experience in guiding companies through ISO 14001 itself. They may have some experience of BS 7750 implementation (or other national environmental systems). They are very useful to an organization that does not have the resources or the knowledge to drive the system forward following the preparatory environmental review. The consultant's task is to use the preparatory environmental review as a technical specification around which to build an effective management system and to assist the client to achieve certification.

Summary

The environmental auditor must be an individual with the correct set of skills. This includes strong interpersonal skills, good time management and a confident manner.

Auditing ability is of course paramount and, although qualifications are necessary in this area, the auditor must have a natural inclination to ask questions about a process or situation until satisfied. The qualities of an auditor must go beyond academic excellence and will include skills developed from auditing a wide spectrum of organizations. Technical experts within the team may assist in this broadening of expertise. Being able to identify environmental issues and have a good understanding of the legal framework is also a fundamental requirement. Conflicts of interest may occasionally occur and the ability to make the correct decision without further reference is also of importance.

Auditors are faced with a large amount of information to analyse during an audit. Some of this information will have been supplied by the client's personnel. Such personnel may give answers that conflict with other answers from other personnel. The dynamics and culture of the audited organization may not be familiar to them and yet, in a relatively short space of time during the audit, they have to make a judgement as to whether the requirements of ISO 14001 are satisfied.

The evolution of the hybrid auditor, someone who is both an environmental expert and a system auditor, was suggested as being likely. With the development of other management systems running in parallel with ISO 14001, the idea of the multi-functional auditor evolving was also suggested.

A brief mention was made about environmental consultants, because some organizations would not be able to obtain certification without expert assistance. Just as there are two types of auditor, a potential client must recognize that there are two types of consultant and make sure the right type is chosen. Certainly asking about the consultant's track record of successfully taking clients through to ISO 14001 is a must. Consultants have a duty to design appropriate systems so that the criticisms that were often levelled at ISO 9000 are never repeated for ISO 14001 systems. These criticisms include:

- Bureaucratic paperwork systems obscuring the fundamental purpose of an ISO 9000 system

- Lack of focus upon quality improvement

- Forcing suppliers to also obtain ISO 9000 – when it may not be appropriate for them

Chapters 2, 3 and 4 include discussions of such criticisms.

It is also worth noting that although the environmental certification process is subject to a framework of controls (as described in Chapter 4), no comparable process of such rigour operates for environmental consultancy.

Chapter 8

EMAS

Introduction

This chapter is primarily intended for those organizations who wish to implement EMAS but may be unsure of the similarities and differences between EMAS and ISO 14001. Are two discrete systems required? Are two systems of documentation required? Does one system complement the other? These and other questions will be answered in this chapter but it must be understood that this chapter is included for breadth of information only. This book does not offer extensive or detailed implementation advice. However, the differences and similarities between the two environmental Standards are highlighted. References to further reading and information sources will be found in Appendix II.

It must also be said that if an organization has understood the concepts of ISO 14001, then implementation of EMAS should present no difficulties or problems as the same fundamental methodology and terminology is used by both standards. The approach taken in this chapter is that of assuming that an organization has already achieved either BS 7750 or ISO 14001, and now wishes to obtain EMAS. This approach is certainly the prevalent one in all the organizations that are registered to EMAS to date.

Just as Chapter 1 described the origins and concepts of ISO 14001, this chapter begins with a description of the birth and concepts of EMAS.

History and concepts of EMAS

Reasons for seeking EMAS certification

EMAS is the initials of the 'Eco-Management Audit Scheme' which is a European environmental management systems standard. It is essentially European Council Regulation (EEC) No. 1836/93 'allowing voluntary participation by companies in the industrial sector in a Community Eco-Management and Audit Scheme'. It was published in its entirety in Official Journal L168 dated 10 July 1993 and was formally launched in the UK in April 1995.

Participating in this programme entitles an organization to register a site on a European Union authorized list of participating sites and to use a European Union approved statement of participation and graphic to publicize inclusion in the programme. Strictly speaking, an organization 'registers' its site to EMAS rather than seeking certification to EMAS as in ISO 14001.

The reason behind achieving this particular environmental management system is that participating organizations can say very publicly that they have nothing to hide regarding environmental issues. They are stating that they do have impacts on the environment but are taking positive steps to reduce such impacts. A great deal of information as to how this will be done must be publicly available. There must be a focus on the provision of numerical data related to raw material and energy usage. By-products and waste products must be described. Certainly the practice of publishing such information pro-actively is better than allowing bad publicity to distort the real situation.

In this way, the organization has the opportunity to put its case first and demonstrate, in a very public way, its achievements in environmental issues.

The European Council hope that the publication of such detailed information will induce companies not just to achieve compliance with the law but to go well beyond.

EMAS is due to be reviewed in the near future by the Commission. This review will include looking at the numbers of EMAS registrations to date. EMAS was intended as a 'self-regulatory' voluntary scheme. However, it could be that if there are insufficient registrations then it may become a compulsory scheme.

In Germany, for example, there are many organizations – chemical companies particularly – registered to EMAS rather than ISO 14001. As a production location, Germany suffers from over-regulation in the fields of environmental protection and safety. EMAS is the first voluntary system enabling organizations to manage environmental protection issues themselves. It is expected, as EMAS registrations spread, that the authorities will ease reporting obligations etc. for those who have implemented EMAS. There are indications of this occurring. If such concessions are not granted, EMAS could begin to be seen as an additional cost burden and a hindrance to further growth. The deregulation aspect is certainly a deciding factor to register for EMAS. Other reasons abound for similar organizations implementing ISO 14001 but with a greater emphasis on the regulatory requirements. The lack of international recognition – EMAS is a European Standard – does not appear to be an issue for some of these organizations.

Relationship with ISO 14001

Because EMAS was developed in parallel with, and at approximately the same time as, BS 7750 the environmental management system requirements of EMAS were met by BS 7750 requirements. Thus an organization that had BS 7750 would comply with most of the requirements of EMAS. BS 7750 is now being replaced by ISO 14001 and this is causing difficulties for some organizations.

Because there are differences between BS 7750 and ISO 14001, organizations that had gained BS 7750 were required to address these differences – 'bridging the gap' – before being awarded ISO 14001. In practice, this was performed at a surveillance visit, usually by a review of documentation. In the UK, for example, UKAS produced an 18-point checklist of areas requiring attention in order to 'bridge the gap' and allow transition from BS 7750 to ISO 14001. (The 18-point check list is included in Appendix II.)

However, ISO 14001 was still not recognized as satisfying the environmental management system requirements of EMAS and further bridging was required to ensure that ISO 14001 could be used as the management systems element. This was an unsatisfactory state of affairs, which caused some concern in interested organizations. As a result of this concern a bridging document,

CEN/PC7/WG EMAS, was produced. The details of this document are covered later in this chapter.

Implementation of EMAS

As indicated in the introduction, only broad guidance is offered by indicating the main steps in the process of implementation. If an organization seeks to implement EMAS, further reading and detailed steps are available from the national Competent Authority for each country within Europe.

EMAS is generally applicable to industrial sites (with the addition of local government in the UK, where district councils have played a key role in maintaining and improving the environment in which their citizens live and work). The environmental statement which must be produced is normally applicable to a single site.

In the UK, local authorities are registered service by service rather than site by site, with an undertaking that the whole authority will be registered by a target date. This requirement for the progressive registration of services is to prevent EMAS-registered 'services' being hampered in their environmental activities by other local authority services. These other services may not be EMAS-registered and could control environmental activities beyond the reach of the EMAS registration services. This is to prevent, for example, an implementing site stating that they cannot control or influence training because another site controls this.

The following broad sectors of industry can apply for EMAS registration:

- Mining

- Manufacturing

- Power

- Waste disposal

A complete list of participating sites (and local authority operations in the UK) is held by the Competent Body in each member state, who are responsible for running the scheme. For example, in the UK it is the Department of the Environment.

Therefore, for an organization to successfully implement EMAS the following seven steps need to be considered:

1   **Develop a corporate environmental policy for the whole organization,** which commits it to compliance with existing legislation and to reasonable, continuous improvements of environmental performance.

2   **Perform a site environmental review (PER),** which covers all aspects of the industrial site or local authority service to be registered.

3   **Develop an environmental programme** that sets quantified objectives and targets for each site.

4   **Implement an environmental management system.** The organization may develop its own system or use a recognized standard – ISO 14001 (but see *Bridging the gap between ISO 14001 and EMAS* below).

5   **Carry out an environmental audit cycle.** The environmental system is checked by the internal auditors to ensure that planned events take place and that documented procedures are being complied with.

6   **Produce an environmental statement.** Each site has to produce a concise and comprehensive statement on its performance to enable the public to understand the environmental impacts of the site. It must be updated annually.

     A full statement must be prepared every three years (maximum). If this option is taken, an interim annual (simplified) statement must also be produced.

7   **Achieve external validation.** The organization's compliance with the terms of the scheme require independent verification.

It will be recognized by an organization that has implemented or has been certified to ISO 14001, that steps 1 to 5 above will have been addressed under the requirements of ISO 14001 anyway, and need not be repeated. Thus the only additional requirement are steps 6 and 7 – the publication of the environment statement followed by an independent external validation by a certification body accredited to verify against EMAS.

The next section will focus on what is required within an environmental statement.

Environmental statements – structure

EMAS is all about organizations taking responsibility for their impacts upon the environment, being committed to continual improvement in environmental performance, and reporting that performance publicly. The environmental statement is the means by which an organization can communicate publicly its progress in managing and improving the environmental impacts of the site's activities.

To register under EMAS, the organization must establish an appropriate mechanism for reporting its progress and achievements.

In essence, the environmental statement provides the organization with the opportunity to produce a balanced account of the overall aims, objectives, targets and other influences affecting the site's overall performance. The public statement will be held with the national regulatory bodies and will be available locally to anyone who wishes to view it.

The information within the statement needs to be understandable. It must set out the nature of the business, the environmental performance it is trying to achieve and the progress it is making. The content of the statement is specified in the EMAS regulation minimum requirements. Just as in ISO 14001 the key word and concept throughout was 'significance', the key word here is 'balance'. Providing the public with information about the site's environmental performance should not be seen as only confessing to failures or only focusing on success.

The format of the statement does not have to be of any specified length or presentation style. EMAS does require that the statement should be able to be read and understood by the public. This implies the minimum use of 'jargon' and technical terms which may be hard for a member of the public to comprehend. Any technical material might be included in an attached appendix. One registered publishing organization produces their statement on glossy art paper illustrated with photographs. This is appropriate and practical for them to produce. A smaller organization with limited resources would be wiser to opt for a simpler approach. However, the technical content must be of a certain minimum standard as outlined below.

The requirement for the public to understand such technical content may, at first sight, represent a significant challenge for the organization. The technical processes of the organization may be very complex and may not be easily described in terms familiar to all members of the public. The public concerned represents a wide cross-section of society, with equally wide levels of understanding of technical issues.

Although the stakeholders will have an interest in the contents of the environmental statement, the 'public' for which the statement is also intended needs some explanation. The term 'public' will be different for larger and multinational organizations compared to smaller, locally-based organizations The target public for an SME may well be only the local public – passers by, local interest groups etc.

Larger organizations with a high public profile may have to cater for the more sophisticated members of the public. Such readers of the statement will require a reasonable amount of technical detail to gain an in-depth understanding of the environmental issues. An environmental statement should contain data and information that is reliable, for it will be subject to verification. The statement should also cover all the significant environmental issues of relevance to the site. It is also necessary to define the basis on which issues are omitted from the statement. If estimates are used then the process of estimation should be described and shown to be technically valid. There are also many ways in which the statements can be made public, rather like the environmental policy requirement of ISO 14001: local and national press, site reception, libraries, on the Internet, etc.

The statement itself should be prepared and made available as soon as the environmental management system is in place and the audit programme has commenced. Subsequent statements should be produced on completion of each audit cycle.

The regulations specify minimum requirements (in Article 5 of the Regulation) and should typically follow a standardized format as follows:

Description of the company's activities at the site concerned
The essential requirement here is that, as with the rest of the statement, the description of the site's activities should be understandable. The use of technical terms should be kept to a minimum and industry jargon avoided. An overview of the manufacturing processes, local environs and a list of products – to assist the reader – should also be included.

An assessment of all the significant environmental issues of relevance to the activities concerned, with aspects classified into major or minor or lesser significance

As well as identifying and describing the relevant environmental issues in non-technical language, there needs to be an accompanying explanation as to how and why these were assessed as being significant.

Any significant changes since the last statement and a deadline for the next statement with targets set and quantified, should also be included.

The company environmental policy, programme and management system pertaining to the site

This should explain what the management at the site is trying to achieve, and how. This should be both in general terms (for example, the policy aims and objectives) and in specific terms (the programme of actions, targets and deadlines). The policy, programme, objectives and targets should relate to and address the significant aspects identified above.

A summary of the figures on pollutant emissions, waste generation, consumption of raw materials, energy and water, noise and other significant environmental aspects, as appropriate

This is not intended to be a mere list of pollutant emissions and quantities of materials used but rather a summary presented in a way which can be easily read, interpreted and conclusions easily reached. Grouping emissions into their environmental impact is one way. Any trends to date should be indicated.

Performance figures should be normalized to production output. This is to negate the distortion created if, for example, a company could demonstrate that waste fell by 50% in a period of 12 months. This would be a fine achievement but meaningless if it had been achieved by halving production output. Thus a balanced set of figures must show details of toxic and hazardous waste, use of utilities, any recycling efforts, emissions to air, discharges to sewer, raw materials usage etc.

A summary of quantitative data on pollutant emissions, waste generation, raw materials, energy and water consumption and other significant environmental aspects (as appropriate) should be included.

Any incidents or breaches of discharge consents must also be included. It is worth noting that a minimum requirement for EMAS is that the site can demonstrate compliance with legal requirements.

Other factors regarding environmental performance
This is an opportunity for the implementing organization to present other issues of environmental importance within the statement. Perhaps newer technologies will be available in the near future – technologies which the organization will use to reduce atmospheric emissions, for example.

The deadline for submission of the next statement
It is essential to state the deadline when the next statement is due – stating the month and the year in an unambiguous fashion.

The identity of the accredited verifier
The name of the accredited verifier must be included on the statement submitted to the Competent Body. Copies of the statements intended to be made publicly available must also include this name. The verifier is the certification body who confirms that the content of the statement is a true record of the state of affairs on the site, and evaluates the performance of the environmental protection system.

ISO 14001 certification bodies can also act as verifiers providing that they conform to EMAS specific accreditation criteria. Individual EMAS auditors are called verifiers. The audit work they perform is very similar to the work of ISO 14001 auditors. However, they are required to have much more in-depth knowledge of environmental legislation and a minimum amount of experience in the environmental field. Such experience covers a broad range of environmental issues.

Bridging the gap between ISO 14001 and EMAS

A problem began to emerge when some ISO 14001 certified organizations considered registering their sites to EMAS. As discussed before, EMAS 'recognizes' BS 7750, but not ISO 14001, as meeting its environmental management system requirements. A European working group was established (CEN/PC7/WG EMAS) to develop a technical comparison of the requirements of the EMAS regulation against the clauses contained in ISO 14001, 14010, 14011, 14012. A bridging document was published as a CEN Technical Report and the significant differences are described below.

Environmental policy

Both ISO 14001 and EMAS required the establishment and maintenance of environmental policies that are committed to continual improvement. However, ISO 14001 does not specifically require an organization to demonstrate EVABAT (Economically Viable Application of Best Available Technology – see Appendix I) whilst EMAS does. An ISO 14001 implementing organization therefore needs to demonstrate EVABAT objectives and evidence of this would be required by the EMAS verifier.

Internal audits

Environmental audit
EMAS refers to 'Environmental Audit' whereas ISO 14001 uses the term 'EMS' audit (meaning environmental management system).

It could therefore be argued that ISO 14001 does not explicitly require that environmental performance is included in the environmental management system audit. However, the 'spirit' of the Standard is about environmental impact control and minimization. Accreditation guidance requires that the principle of continual improvement of environmental performance is apparent and other clauses of ISO 14001 do much to promote the intended spirit of the Standard.

Thus an organization that has obtained ISO 14001 certification would be performing internal audits of the correct methodology (for example, effects-based auditing). It may be that the organization's audit protocols may need to be reviewed to reflect this emphasis.

Audit cycle
EMAS states that the audit cycle should be completed at intervals no longer than three years. ISO 14001 states the requirement for establishing and maintaining a programme of audits – but no frequency is stated.

ISO 14001 organizations will have derived an audit programme cycle to audit the most significant areas of operation first – perhaps both compliance and effects-based audits. Such a programme may spread over twelve months and may not cover all activities. Areas of operation deemed to be insignificant may well not be audited for some time. Thus such organizations wishing to register to EMAS would have to ensure that the audit programme would capture all areas of activity within three years, and would be documented accordingly.

However, it must also be remembered that the ISO 14001 auditor, during routine six-monthly surveillance visits of ISO 14001 registered organizations, will always review the internal audit function. The auditor would, in all probability, ask why an area or function had not been audited, for example, within 12-18 months of certification.

Environmental review

EMAS states organizations wishing to be registered must have undertaken a preliminary environmental review. ISO 14001 only requires such a review if there is no previous documentation of environmental activities.

An ISO 14001 auditor when performing the pre-audit will look for evidence that the organization has performed some environmental background research and study. On what have they based their system? How did they decide what was important and significant? In practical terms, the organization will have carried out an environmental review (the PER) and this may well be on differing levels of formality, complexity and scope (depending upon the organization's environmental aspects and the resources available to carry out such a review). The auditor cannot assess or judge the review itself in isolation, as this is beyond the scope of ISO 14001. However, unless an organization can demonstrate the reasons for the way it has implemented the front end of its environmental management system, the auditor can expect that the rest of the system will be fundamentally flawed.

Indeed, operating experience has shown that organizations implementing a system based upon subjective evaluations experience major difficulties. The environmental policy is then based upon these subjective evaluations and an environmental management system designed around this policy. Sooner, rather than later, the system collapses and the organization has to start again – with losses of both time and money.

Therefore, in the unlikely event that an organization had not undertaken a reasonably structured and searching environmental review, they would need to perform this prior to EMAS verification.

These are the main 'gaps' between what ISO 14001 offers and what EMAS requires. Further areas that organizations are recommended to address are as follows:

- EMAS is very specific about requirements for external communications. ISO 14001 requirements are somewhat more diffused and some review of documentation and procedures may be required.

- EMAS is quite specific about contractors coming on-site – and their environmental obligations. ISO 14001 addresses such requirements via 'operational controls' and 'training, awareness and competence'.

- EMAS uses the term (with respect to compliance with environmental legislation and regulations) 'provision for compliance'. ISO 14001 uses the term 'commitment to comply' and this may be seen to imply less importance. An organization would need to demonstrate provisions for compliance before EMAS registration.

In conclusion, the spirit and intention of the two Environmental Standards are the same. Therefore, any organization certified to ISO 14001 and wishing to register for EMAS will find that their system fundamentally meets the environmental management system requirement. However, they may have to change or amplify some elements of their documented environmental management system. Unfortunately, at the time of writing of this book, debate is still in progress as to whether ISO 14001 fulfilled all of the requirements and components of EMAS. Although radical new decisions and changes are not anticipated, some minor alterations are to be expected.

The bridging document, as outlined above, is not a specification against which external auditing can be performed (with subsequent registration). A possible solution for the organization is to conduct an internal audit – following the bridging document guidance – during implementation of ISO 14001. Documentary evidence of such guidance (in the internal audit report, for example) can be used as objective evidence, during the EMAS verification audit, to demonstrate to the verifier that the 'gap' has been bridged.

EMAS or ISO 14001? A choice for organizations

An organization has three choices when it comes to deciding whether to go ahead and implement an environmental management system. It can choose one of the following:

1 Obtain ISO 14001 initially in its own right and then, after a suitable period of time (dictated by the organization itself), register for EMAS.

 The fact that the organization has an existing environmental management system is taken fully into account during the EMAS verification to the point that additional verification of the system is not performed (that is, part of the requirements of EMAS have been met). The verification will then focus very much on validation of the environmental statement. (However, this route depends very much on the outcome of the above discussions regarding the bridging process.)

2   Be assessed to ISO 14001 and verified to EMAS at the same time.

EMAS verification and ISO 14001 assessment can be performed concurrently by a certification body using a team which includes an EMAS verifier.

Failure to reach some of the requirements of EMAS (in the environmental statement) may mean that certification to ISO 14001 is allowed but there is a delay in obtaining EMAS until corrective actions have been taken.

Failure to meet the requirements of ISO 14001 will delay both certifications as all the elements of ISO 14001 need to be in place to achieve EMAS. Again, one is mindful of bridging discussions.

3   Register only to EMAS.

The verification team will of course look at all the elements required of an environmental management system (that is, ISO 14001) but should the audit be successful, only EMAS registration is granted.

Obviously, an organization achieving both standards not only has to produce the requisite environmental statement at the correct times but is also subject to ongoing six-monthly surveillance visits in order to comply with the requirements of ISO 14001.

When successfully verified by the external verifier, the company will be awarded the ECO Audit Logo which must always be accompanied by a schedule detailing the sites involved. Several alternative phrases and templates are used (as appropriate) and these are described in the Regulation itself.

Summary

EMAS is a European standard and only time will tell whether it will have the same recognition, acceptability and credibility outside Europe as ISO 14001. Both standards have the same underlying principles but EMAS goes a step further in requiring the organization to publish data on its environmental performance. This was seen by the writers of EMAS to be a desirable approach for organizations to demonstrate that they were environmentally responsible and had nothing to hide. It was hoped that the uptake, on a voluntary basis, would be high. Some organizations, however, saw EMAS as a threat – reasoning that publishing such data and thereby revealing commercially sensitive information about their processes might give their competitors an unfair advantage.

ISO 14001 has strong links with EMAS, as was suggested in the preceding paragraph, because they have common underlying environmental principles. However, there is still some debate as to whether ISO 14001 satisfies all of the management systems requirements of EMAS.

EMAS has implementation parallels with ISO 14001 but operates only in specific sectors. The concept of the environmental statement as a document open to the general public has been discussed. The idea of a 'general public' was questioned and was refined, depending very much upon the public exposure, complexity and size of the organization.

However, there is a problem for those organizations who have ISO 14001 and wish to use it as the management system element of EMAS. Due to several factors in the drafting processes, there is a gap which needs to be bridged before it can be said that ISO 14001 fulfils the environmental management system requirement of EMAS.

An organization choosing between implementing ISO 14001 and EMAS has to consider which of these standards is more widely recognized. ISO 14001 is recognized world-wide whereas EMAS is not very well known outside Europe and has yet to be recognized internationally as an acceptable alternative to ISO 14001.

Appendix I

Glossary of terms

Accreditation

A process whereby a certification body is subjected to audit to ensure that it is qualified to issue certificates in certain business sectors based upon its individual auditors' expertise, background, training and qualifications.

A rigorous assessment of the certification body's head office procedures is carried out (for example, ensuring that reviews of audit team selection process are carried out). Some certification audits may also be 'witnessed' by UKAS (in the UK). Continual surveillance is also part of the monitoring process.

In the UK and elsewhere, non-accredited certification bodies can operate legally. However, there is no higher authority that they are responsible to, and in some instances the value of such certification can be worthless.

In the UK when a certificate is issued by an accredited body the certificate bears the 'Crown and Tick'. (See *Crown and Tick*.) This is a logo bearing an image of the royal crown coupled with a large 'tick', and is a respected mark of certification integrity not only in the UK but world-wide.

Agenda 21

One of the outputs of the Earth Summit in June 1992 was a report entitled 'Agenda 21' – a large document of some 40 chapters, all connected to environmental issues, covering for example:

- Changing consumption patterns

- Integrating environment and development into decision making

- Promoting education, public awareness and training

It is not a legally binding document but sets out a global consensus addressing the pressing environmental problems of today and aims at preparing the world for the challenge of the next century. As such it places great emphasis on the need for all sections of society to participate in working towards sustainable development, including business and industry.

Assessment process

The whole process of certification performed by the third party certification body. This includes the pre-audit, desk study and the certification audit followed by the issue of a report and certificate.

Authorized process

In the UK certain industrial processes are required to be authorized. The organization must apply to either the Local Authority for part B processes (emissions to air only processes) or the Environment Agency for part A processes (emissions to all three media).

Such emissions need to be monitored and the results of such monitoring are sent to the appropriate authority. The results are also kept in the form of environmental records on site, available for inspection at any time. Breaches of such authorization must be notified and actions must be taken to minimize such breaches in the future. The Authority or Agency may require certain improvements to be completed by a certain date. Prosecutions can follow in cases of persistent offenders.

Across Europe similar conditions apply although details may be different due to national interpretations of European Directives.

BATNEEC

A complex definition arising out of the Environmental Protection Act 1990 – an organization must aim to control emissions using the Best Available Techniques Not Entailing Excessive Cost. The intention is to focus the organization's management into making reasoned decisions about the technology used to control emissions, without forcing financially crippling penalties.

BPEO

A phrase encompassed within the Environmental Protection Act 1990. This is a requirement to minimize pollution by applying the Best Practical Environmental Option. BPEO requires that the environmental implications of all the disposal options available is evaluated, and the option chosen results in the least environmental damage and is consistent with the prevailing regulations.

Brownfield site

A site that is developed for industry or domestic use that has had previous use. The positive environmental impact is that such use of derelict land preserves more virgin or greenfield sites. Obviously this is more important in smaller countries which do not have unlimited greenfield sites.

Certification

The process of issuing a certificate to an organization that has achieved compliance to a recognized standard and has been independently audited by a third party to such a standard. Such a certificate can then be used as evidence by the organization to demonstrate compliance to any interested party. In the context of this book this is compliance and certification to ISO 14001.

Certification bodies

Independent organizations whose sole business is auditing other organizations' quality and environmental management systems for compliance against national and international standards. Most of the bodies originally started assessing to the standard BS 5750 (now ISO 9000) and have extended their scope into environmental certification. They have no commercial interests and must be seen to be totally independent so that there can be no question of bias when issuing a certificate.

It is not sufficient in the world of business for a organization to state 'we are environmentally responsible and have a system which conforms to ISO 14001'. This statement must be independently audited by the certification body.

The award of the certificate is of tremendous value to the client. It is not obtained easily and is subject to continuous surveillance visits to ensure compliance. Accredited certification bodies themselves are audited by the national accreditation body (for example UKAS in the United Kingdom), to ensure compliance with higher level 'standards'.

Corrective action requests

During the certification audit, or during routine surveillance visits, by the certification body, if the external auditor discovers areas of concern, or nonconformity, which may jeopardize the integrity of the environmental management system, a corrective action request will be generated by the auditor. This corrective action request will either be classed as major or minor, depending upon the severity of the nonconformance. For a major corrective action request, a time-scale of one to two months is given to the client for it to be satisfactorily addressed. This usually necessitates an extra surveillance visit by the auditor to observe evidence that corrective action by the client has taken place. Minor corrective actions requests are usually allowed a time-scale of six months. They are verified by the auditor by being addressed at the next planned surveillance visit (see also *Noncompliances*).

Crown and Tick

In the UK, for historical reasons, the sign of accredited certification used by UKAS is a logo bearing an impression of the Royal Crown plus a 'tick'. However, the use of that mark suggests Royal Assent (that is, royal permission or approval). Organizations that achieve accredited certification have no such assent; thus restrictions are placed upon them.

In the UK, organizations cannot use the UKAS mark on flags, vans or trucks, primary packaging and promotional goods such as diaries. Elsewhere, national accreditation bodies may use their national accreditation mark in many of these instances without restrictions.

DIS (Draft International Standards)

International standards go through several stages when being written and the draft stage is reached near to the first publication of the final Standard itself. Any differences between the draft and final standards tend to be of minor or even just typographical changes. For example one of the changes between DIS/ISO 14001 and ISO 14001:1996 was that the clauses were numbered differently. Due to consumer demand, and with the national accreditation body agreement, organizations can be certified to a draft standard.

Direct aspects/impacts

See *Environmental aspects*.

Discharge consent

This can be considered as a contract between the organization and the local water authority whereby they permit the organization to discharge effluent providing that it meets certain parameters (such as minimum or maximum pH, maximum levels of heavy metals, volume and rate of discharge). The authority may require the organization to monitor and log such discharges and either send in the results monthly or six monthly, or they may simply require them to be available for inspection by their inspector. If such limits are exceeded, the organization is obliged to inform the authority as to what corrective actions were taken. The authority will possibly charge a fee for treatment of the excess effluent burden and, in persistent or extreme cases, will prosecute the organization in a court of law.

DTI (Department of Trade and Industry)

In the UK, a branch of the Government responsible for a diversity of activities which includes monitoring the activities of the national accreditation body, UKAS.

EMS (Environmental management system)

A management system that enables an organization to control its impacts on the environment. It may well be an informal or fragmented system based on perceived impacts or driven purely by the requirements of legislation. However, the term as understood and discussed in this book means a management system that not only controls its impacts, but takes reasoned and logical steps to minimize environmental impacts and uses the tools of measurement and monitoring. Such a system needs to have a certain minimum level of documentation so that it can be followed by personnel and internal and external auditors.

Environmental aspects

An element of an organization's activities, products or services that can interact with the environment.

Environmental impact

Any change to the environment, whether adverse or beneficial, wholly or partially resulting from an organization's activities, products or services.

An environmental effect has two components:

1   Aspects – activities which can have a beneficial or an adverse effect.

2   Impacts – which take place as a result of an aspect.

There are two categories of the above:

1   Direct impacts are those impacts that an organization can directly control.

2   Indirect impacts are those impacts that an organization can only influence by various means. In practice, this means customers and suppliers.

Environmental audit

A management tool comprising a systematic, documented, periodic and objective evaluation of how well an organization's management and equipment are performing, with the aim of contributing to safeguard the environment by:

• Facilitating management control of environmental practices

• Assessing compliance with organization policies and regulatory requirements

Although there are several types of environmental audit, suited for different purposes, the above definition related to compliance with ISO 14001 is sufficient.

Environmental verifiers

In EMAS language, this is the equivalent of the third-party ISO 14001 auditor. There are however more requirements placed upon the individual as regards environmental training, experience and competence.

End-of-pipe technology

The application of resources and technology almost 'after the event' is considered to be 'end-of-pipe technology'. As an example, in order to reduce a pollution incident, such as a spillage of acid, automatic alkali-dosing equipment is employed. This will make the consequences of an incident less serious by application of the technology but the resources and technology should in fact be used at the front end of the process to ensure that likelihood of a spillage is minimized.

EAC (European Accreditation of Certification)

Within Europe, accreditation bodies are anxious to adopt a coherent approach to certification for ISO 14001 and collectively have formed the EAC. Hopefully, this will be further widened to include Asia and North America to bring global harmonization of certification processes, allowing reciprocal agreements on acceptance of national standards and mutual recognition of national accreditation bodies.

EVABAT

Economically Viable Application of Best Available Technology. This is another acronym for the use of technology to control environmental impacts without causing an organization financial hardship.

Greenfield site

A site that is developed from virgin land. This has many advantages – it may be in an unspoilt area to attract the workforce and tax/grant allowances. However, it is using up valuable land area.

Indirect environmental aspects/impacts

See *Environmental impacts*.

IPC (Integrated Pollution Control)

Relates to the controls exerted by an organization on its impacts on the environment as a whole, over all three media: air, land and water. IPC ensures that solving one environmental problem does not create another elsewhere.

IPPC (Integrated Pollution Prevention Control)

EC Directive 96/61/EC (24/10/96). IPPC will be a licence to operate industrial processes. To gain IPPC an organization needs to meet the following summarized requirements:

1   All the appropriate preventive measures are taken against pollution.

2   No significant pollution is caused.

3   Waste production is avoided. Where waste is produced, it is recovered. Where this is technically and economically impossible, it is disposed of while avoiding or reducing any impact on the environment.

4   Energy is used efficiently.

5   The necessary measures are taken to prevent accidents and limit their consequences.

6   The necessary measures are taken to avoid any pollution risk once the business activities on the site have ceased.

New installations will require authorization prior to the start of activities.

Existing installations will need to be authorized under IPPC by 30/10/2004.

Landfill Tax

In the UK a recent tax (1996) which is levied upon specified wastes that are disposed of to landfill, in an effort to persuade such producers of waste to reduce quantities going to landfill by being more efficient and recycling.

It is also worth noting that natural landfill sites such as clay and sand pits and disused quarries are becoming scarcer in the UK and newer ones require considerable and costly engineering. This will result in landfill cost spiralling upwards.

Other European countries have similar taxes to discourage landfill as a solution to waste disposal.

Local Agenda 21

Over two thirds of the Agenda 21 cannot be delivered without the commitment and co-operation of local government. The key role of local authorities is set out in chapter 28 of Agenda 21. Each local authority is encouraged to adopt its individual 'Local Agenda 21' by 1996, with its own sustainable development strategy at the local level involving partnerships with other sectors such as businesses, community and voluntary groups.

Much guidance has been published to assist local authority economic development officers who are trying to reconcile environmental and economic aims against a background of both growing international environmental pressures and severe resource constraints within their authorities. Agenda 21 recognized that action at local level is vital and encouraged local authorities (as the level of Government closest to the people) to develop sustainable development strategies for their communities.

Although not a statutory requirement many UK local authorities are establishing local Agenda 21 programmes – working with the local community to take action to ensure greater sustainability.

Local Agenda 21 involves a process of consultation and consensus between local authorities, citizens, local organizations and business enterprises. It outlines a strategy for the continued economic and social development of the world without detriment to the environment and natural resources and aims to provide guidance for governments in establishing environmental policies that meet the needs of sustainable development.

One of the recommendations of the Earth Summit at Rio was for individual countries to prepare strategies and action plans to implement agreements from the summit conference – the 'Rio Agreements'.

Internationally, local Agenda 21 initiatives are being co-ordinated by the ICLEI (International Council for Local Environmental Initiatives) based in Canada (see Appendix II).

LCA (Life cycle analysis)

Based upon a consideration of all the environmental impacts of a product or system 'from the cradle to the grave' – that is, from raw material extraction and processing through manufacture, distribution and usage to ultimate disposal of the product and waste management. A set of ISO 14000 Standards is currently under consideration (see Appendix III).

Noncompliances

Discoveries during a management systems audit which demonstrate that a documented procedure or work instruction is not being adhered to by the relevant personnel or an objective is not being met through individual targets not being reached. By the process of generating a corrective action request, the 'nonconformance' or 'noncompliance' can be brought to management's attention and, following an investigation, steps can be taken to ensure the same problem does not occur again.

SMS (Safety management system)

A management system built up in a similar fashion to an environmental management system so that occupational health and safety is managed in a structured way and that continual improvement in safety practices is achieved.

Sector Application Guides (SAGs)

These are guidance notes published by different business sectors to assist organizations in that sector to design more easily an environmental management system. For example, the guide may indicate where emphasis should be placed or where, for example, clauses of the Standard may not be totally applicable.

They are published not by one organization but by many and as such there is no consistency of approach or depth and breadth of assistance offered (see Appendix II).

Significant environmental aspects/impacts

A ranking placed upon environment aspects as identified by an organization, as having a greater impact upon the environment than other aspects.

Small to medium enterprises (SMEs)

There is no official definition of what an SME is. However, in Europe when industry statistics are compiled, the definition is based upon employee numbers rather than turnover, profits, market size or number of sites:

Number of employees	Classification of organization
0-9	micro
10-99	small
100-249	medium
250+	large

Therefore organizations with between 10 and 249 employees are SME's.

However, in the context of this book, the definition of an SME is further refined as follows. An SME:

• Will have working managers and perhaps owner who have little or no time available for new projects outside direct business activities.

• Will probably be operating in an informal style and may not have documented management systems.

• Will be deficient in technologically trained personnel (that is, university graduates).

• Cannot afford to employ a single discipline manager such as a quality assurance manager or environmental manager.

• Will undertake 'on the job' training with no structure for identifying training needs.

Therefore, such an organization, although it may be very successful financially, may not have the resources for ISO 14001 implementation.

Sustainable development

Many definitions abound but the 'Brundtland Definition' is perhaps the best one. This is named after the Prime Minister of Norway, Gro Harlem Brundtland, who chaired the UN-sponsored World Commission on Environment and Development (WCED) in its report 'Our Common Future' published in 1987. This definition has its attractions due to its simplicity. Essentially, it confirms that continued economic and social development is vital but that this must be without detriment to natural resources (including air, water, land) and biodiversity (as key resources), especially as continued human activity and further development depend on the quality of these resources. The definition is:

Development which meets the needs of the present without compromising the ability of future generations to meet their own needs.

TC 207

An ISO technical committee (TC 207) was established to produce internationally agreed environmental management systems and related documents, by working through six sub-committees looking at environmental auditing, environmental labelling and life cycle analysis. One of these sub-committees was Sub-committee 1, and is responsible for ISO 14001 development considering three areas:

1   Environmental management systems specification with guidance for use. ·

2   Environmental management system general guidelines and principles and application.

3   Environmental management system guidelines on special considerations affecting small and medium sized organizations (SMEs).

TQM (Total Quality Management)

There are several similar definitions but essentially TQM can be considered to be the way an organization enacts a philosophy of continually striving to meet its customer requirements and exceeding those expectations. It goes far beyond quality assurance and calls upon enhancement of attitudes of all staff in everything they do.

Sometimes TQM is used incorrectly to describe an integrated management system.

Transfer note

When waste is passed from one organization to another, the organization accepting the waste must have a written description of it. It must be signed by representatives of both organizations and contain the following information:

- Description of the waste and quantity

- The time and date the waste was transferred

- Where the transfer took place

- Whether one of the parties is a waste carrier. In the UK and elsewhere, a licence is required from the national equivalent of the Environment Agency to transport waste

- Whether one of the parties has a waste management licence – again authorized by an equivalent of the Environment Agency

The above will be embodied in applicable national legislation with reference to due diligence or duty of care.

UKAS (United Kingdom Accreditation Service)

In the UK, this body was set up by the DTI in 1995 as a replacement to the NACCB to provide national accreditation of certification bodies and operates with similar European, American, Australian and Asian bodies. It is an organization limited by guarantee and its members represent various interests in accreditation.

UKAS operates to international standards that outlaw accredited certification bodies providing both consultancy and certification services to the same organization, in order to maintain unquestionable impartiality in all activities.

Volatile organic compounds (VOCs)

Covers a wide range of organic chemicals which evaporate at ambient temperatures. They include many commonly used industrial solvents such as white spirit, xylene, toluene, acetone and methyl ethyl ketone (MEK) solvents which are important components of products such as adhesives, varnishes, paints and other coatings.

For example, the oil industry is a significant source of VOCs and the high VOC content of petrol means that road traffic is the greatest single source of VOCs in the UK (being responsible for over 40% of total emissions).

Appendix II

Further reading

All addresses in this appendix are in the UK unless specified otherwise.

Standards

Quality assurance management

ISO 9001:1994 'Quality Systems – Model for Quality Assurance in design, development, production, installation and servicing'

ISO 9002:1994 'Quality Systems – Model for quality assurance in production, installation and servicing'

ISO 9003:1994 'Quality Systems – Model for quality assurance in final inspection and test'

ISO 9004:1994 'Quality Systems – Guide to Quality Management and Quality Systems Elements for Services'

Occupational Health and Safety

BS 8800:1996 'Guide to Occupational Health and Safety Management Systems'

Information Security

BS 7799:1995 'British Standard Code of Practice for Information Security Management'

The above standards can be obtained from:

The British Standards Institute
Customer Services
389 Chiswick High Road
London W4 4AL

Addresses/contacts

EARA

Environmental Auditors Registration Association – based in the UK but with international membership, this is a voluntary membership scheme for environmental auditors and has its own courses and examination schemes.

EARA
Fen Road
East Kirby
Lincolnshire PE23 4DB

SCEEMAS

The Small Company Environmental and Energy Assistance Scheme – giving grant assistance to SMEs in the UK for environmental consultancy and implementation.

The Sceemas Office
NIFES Consulting Group
NIFES House
Sinderland Road
Broadheath
Altrincham
Cheshire WA14 5HQ

Reading

Books/General

NHBS Mail-order Bookshop
2-3 Wills Road
Totnes
Devon TQ9 5XN

Stocks over 600 titles of key handbooks, manuals, references, guides and reports covering all environmental issues.

The Eco-Management and Audit Scheme: A Practical Implementation Guide
British Library   ISBN 0 946655

Contains the level of detail necessary to implement EMAS.

Guidelines on the Accreditation of Certification Bodies for Environmental Management Systems, Document EAC/G5

The full text (an outline of which is in Appendix IV). In the UK, obtainable from:

UKAS
Audley House
13 Palace Street
London SW1E 5HS

Environmental Management and Business Strategy
Welford, R and Gouldson, A (1993), Pitman Publishing, London

Background reading on using environmental management as a business tool.

Responsible Care Management – Guidelines for Certification to ISO 9001 – Health, Safety and Environmental Management Systems in the Chemical Industry
Chemical Industries Association ISBN 0 900623 85 3

Additional reading adding to the concepts outlined in Chapter 5.

Periodicals

ENDS (Environmental Data Services Ltd)
Finsbury Business Centre
40 Bowling Green Lane
London EC1R 0NE

A mechanism for keeping up to date with environmental issues from around the world.

Environmental Business (Fortnightly Newsletter)

Available from:

Information for Industry Ltd
18-20 Ridgeway
London SW19 4QN

Reporting of topical environmental issues.

The Warmer Bulletin

Published by:

The World Resource Foundation
Bridge House
High Street
Tonbridge
Kent TN9 1DP

More in the style of 'green' campaigning but a good read for global environmental issues.

Environmental reports

Earth Summit, Agenda 21: the United Nations programme of action from Rio
United Nations, 1993 ISBN 92 1 100509 4

A good, if somewhat long background read on the 'environment' and the challenges facing the world.

Environmental Management Standards – Accelerator or Brake for Business?

A pan-European report conducted by the Institut Superieur de Commerce International a Dunkerque. The aim was to examine what motivated companies to seek one or more of the environmental standards. The survey was based on responses to questionnaires completed by more than 500 companies in France, Germany, the Netherlands and the UK in 1996.

Available from:

> ISCID
> 129 Avenue de la Mer
> BP69
> 59942 Dunkerque Cedex 2
> France

or:

> SGS Yarsley ICS Ltd
> SGS House
> Portland Road
> East Grinstead
> West Sussex RH19 4ET

Small Firms and the Environment: A Groundwork Status Report

Although now becoming dated, written in November 1995, this small report nevertheless highlights that SMEs do have practical difficulties in facing environmental challenges – including implementation of environmental management systems. Obtainable from:

> Groundwork National Office
> 85-87 Cornwall Street
> Birmingham B3 3BY

Integrated Management

Croners Special Report, Issue 15, August 1996
'Occupational Health and Safety and Environmental Management'

Several page report on integration of BS 8800 with ISO 14001.

Croners Special Report, Issue 17, November 1996
'Integrating Environmental and Quality Management'

Several page report on integration of management systems. (The address for Croners appears on the next page.)

ISA 2000 - a standard for occupational health and safety

Guidance document setting out the clauses for this assessable occupational health and safety standard.

Available from:

    SGS Yarsley ICS Ltd
    SGS House
    Portland Road
    East Grinstead
    West Sussex RH19 4ET

Environmental law/guidelines

A Pocket Guide to Environmental Law by David E Shillito

Published by:

    Intelex Press Ltd
    62 Kings Street
    Maidenhead
    Berkshire SL6 1EQ

This is a useful booklet for the environmental auditor to carry around as a summary of legislation relating to air, water and land.

Barbours Health, Safety and Environmental Index

A useful subscription to take out for keeping abreast of legislation.

Available from:

> Barbour Index Plc
> New Drift Road
> Windsor
> Berkshire BL4 4BR

Croners Environmental Management

Includes updating amendment service and bi-monthly briefing notes.

Croners Environmental Policy and Procedures

Written in the style of a workshop manual to be used during the implementation of an environmental management system. Included are pro-formas of registers etc. and documentation likely to be used.

Includes special reports as above in the section on integrated management. Both the above publications focus mainly on UK law. Some European and North American legislation is also referenced. Published by:

> Croners Publications Ltd
> Croner House
> London Road
> Kingston upon Thames
> Surrey KT2 6SR

EMAS

EEC Council Regulation No 1836/93 Community Eco-Management and Audit Scheme. Official Journal of the European Communities No. L168 10 July 1993.

This is essential reading for any organization looking for EMAS registration, and is available in the UK from:

> UK Competent Body
> The EMAS Registration Office
> Department of the Environment
> Sixth Floor
> Ashdown House
> 123 Victoria Street
> London SW1 6DE

Sector Application Guides (SAGS)

Croners Special Report, Issue 14, July 1996

Lists, with addresses, where SAGs can be obtained from in the UK. They are all written around the requirements of BS 7750, and cover the sectors of agriculture, food, textiles, furniture, paper, oil refineries, chemicals, rubber and plastics, electricity transmission, transport, healthcare, waste management, foundries, ceramics. (Address for Croners appears on previous page.)

BS 7750 and ISO 14001

The transition from BS 7750 to ISO 14001

The 18-point 'checklist' produced by UKAS.

An organization that has achieved BS 7750:1994 needs to address the following points in order that its existing system will meet the requirements of ISO 14001:1996. The organization must ensure that:

1   A commitment to the prevention of pollution is included in the environmental policy – clause 4.2(b) 'Environmental Policy'.

2   A commitment to comply with relevant environmental legislation and regulations, and other requirements to which the organization subscribes, is included in the policy – clause 4.2(c) 'Environmental Policy'.

3   The system is designed to take due regard of the requirements for preventive action set out in clause 4.5.2 'Nonconformance and corrective and preventive action' of ISO 14001.

4   The system is designed to control environmental aspects over which the organization can be expected to have an influence in order to determine those aspects which have or can have significant impacts on the environment – clause 4.3.1 'Environmental aspects'.

5   The management review is undertaken by top management – clause 4.6 'Management review'.

6 The EMS audit programme must be based on environmental perform-ance of the activity concerned – clause 4.5.4 'Environmental manage-ment system audit'.

7 Monitoring and measurement must also include recording of informa-tion to track performance, relevant controls and conformance with the organization's objectives and targets – clause 4.5.1 'Monitoring and measurement'.

8 Monitoring and measurement must also include establishment and maintenance of a documented procedure for periodically evaluating compliance with relevant environmental legislation and regulation – clause 4.5.1 'Monitoring and measurement'.

9 The system must provide that the organization consider processes for external communication on its significant environmental aspects and record its decision – clause 4.4.3 'Communication'.

10 Training must include awareness of the environmental benefits of improved personnel performance – clause 4.4.2(b) 'Training, aware-ness and competence'.

11 Training must include awareness of roles and responsibilities including emergency preparedness and response requirements – clause 4.4.2 (c) 'Training, awareness and competence'.

12 The management representative(s) must have a defined role, responsi-bility and authority for reporting on performance of the environmental management system to top management for review and as a basis for improvement of the environmental management system – clause 4.4.1 'Structure and responsibility'.

13 Objectives and targets should be consistent with the environmental policy and include the commitment to prevention of pollution – clause 4.4.3 'Objectives and targets'.

14 Regard must be had to the concepts as defined (clauses 3.3 and 3.4) of environmental aspects and environmental impacts – clause 4.3.1 'En-vironmental aspects'.

15 The environmental policy must support the organization's intentions and principles in relation to its overall environmental performance

which provides a framework for action – clause 3.9 'Environmental Policy'.

16 Continual improvement is not qualified by the EVABAT limitation in BS 7750.

17 The environmental performance to be achieved by the management system must be defined in line with clause 3.8 'Environmental performance'.

18 Emergency preparedness and response procedures must include provision for preventing and mitigating the environmental impacts of accidents and emergency situations, as well as identifying the potential for such situations – clause 4.4.7 'Emergency preparedness and response'.

Appendix III

ISO 14000 series of standards

All the following standards can be obtained from:

The British Standards Institute
Customer Services
389 Chiswick High Road
London W4 4AL
UK

ISO 14004:1996 'Environmental Management Systems – General Guidelines on Principles, Systems and Supporting Techniques'

This has been developed to provide additional guidance for organizations on the design, development and maintenance of an environmental management system. It is not intended to be certified against. It is for those organizations who may feel that they require some additional guidance and background information on the underlying principles and techniques necessary to develop such a system.

These include:

* Internationally accepted principles of environmental management and their application to the development of an environmental system

* Practical examples of issues arising during the design of the system

* Practical help sections on system design, development, implementation and maintenance

ISO 14010:1996 'Guidelines for Environmental Auditing – General Principles on Environmental Auditing'

A generic environmental auditing standard which sets down guidelines on the general principles involved in environmental auditing.

ISO 14011:1996 'Guidelines for Environmental Auditing – Audit Procedures – Auditing of Environmental management Systems'

Provides guidance on the audit procedures required in order to plan and conduct an environmental management systems audit.

ISO 14012: 1996 'Guidelines for Environmental Auditing – Qualification Criteria for Environmental Auditors'

Sets out the minimum qualification criteria for environmental management system auditors and lead auditors. It also provides indicators to employers and clients on how to evaluate the suitability of auditors.

Other environmental standards

(Some of these have not yet been published.)

ISO 14002 'Environmental Management Systems'

Guidelines on special considerations affecting small and medium enterprises. At preliminary stage – evaluating market need.

ISO 14014 'Initial Reviews'

Being considered for inclusion in 1999 revisions of ISO 14001 and ISO 14004.

ISO 14015 'Environmental Site Assessments'

Preliminary stage. Market need being evaluated.

ISO 14020 'Environmental Labels and Declarations – General Principles'

Provides guidelines and principles for self-declaration environmental claims (Environmental Labelling) made by manufacturers, importers, distributors and retailers of products.

ISO 14021 'Environmental Labels and Declarations – Environmental Labelling – Self Declaration Environmental Claims – Terms and Definitions'

Guidance on the use of terms for self-declared environmental claims.

ISO 14022 'Environmental Labels and Declarations – Environmental Labelling – Self Declaration Environmental Claims – Symbols'

Guidance on self-declared environmental claims when using symbols.

ISO 14023 'Environmental Labels and Declarations – Environmental Labelling – Self Declaration Environmental Claims – Testing and Verification Methodologies'

Guidance on testing and verification of self-declared environmental claims.

ISO 14024 'Environmental Labels and Declarations – Environmental Labelling – Type I – Guiding Principles and Procedures'

Guidance for establishing a certification programme for third-party environmental claims.

ISO 14025 'Environmental Labels and Declarations – Environmental Labelling – Type III – Guiding Principles and Procedures

Guidance on profiling of product environmental effects.

ISO 14031 'Environmental Performance Evaluation – Guidelines'

Provides guidelines and principles for determining environmental performance of an organization.

ISO 14040 'Life Cycle Assessment – Principles and Framework'

Principles for carrying out and reporting of LCA studies.

ISO 14041 'Life Cycle Assessment – Life Cycle Inventory Analysis'

Methodology for definition of goal and scope, performance of LCA, interpretation and reporting.

ISO 14042 'Life Cycle Assessment – Impact Assessment'

Provides guidelines and principles for determining environmental impacts arising from the production, use and disposal of a product or provision of a service.

ISO 14043 'Life Cycle Assessment – Interpretation'

At working draft stage.

ISO 14050 'Environmental Management – Terms and Definitions'

Provides terms and definitions for the above standards.

Appendix IV

Guidelines for the accreditation of certification bodies for environmental management systems

EAC/G5, published in June 1996

This guidance document brings together the national accreditation systems of 17 European countries and is available in the UK from UKAS (address in Appendix II).

The guidelines are based on (and modified where appropriate) the European Standard EN 45012 *'General Criteria for Certification Bodies operating Quality System Certification'*, i.e. the ISO 9000 series of standards. They comprise 19 clauses and consist of guidance primarily for certification bodies. However, from reading this document, implementing organizations can gain some understanding of the way certification bodies are required to operate and some of the reasons for the methodology of preparation for, and performing of, the assessment process. An understanding should also be gained of how accredited certification is highly regulated to ensure that such certification is meaningful and respected.

A brief description of each clause follows, with a fuller explanation of those clauses which are likely to affect an implementing organization – and these are identified in italics.

Clause 1

Defines the object and field of application of the guidelines – and states that it is intended for the use of bodies concerned with recognizing the competence of certification bodies (i.e. accreditation bodies).

Clause 2

Lists a set of definitions concerning standardization in this field.

The coverage of the ISO 14001 certificate is described in terms of the location of the organization to be certified (i.e. the site) and the requirements where an organization has multiple sites, such as sampling methodology. Shared facilities on one site are also described.

Clause 3

General requirements of the certification body, including a directive that all organizations shall have access to the services of the certification body. There must be no hidden discrimination either by speeding up or delaying application for certification.

Clause 4

The administrative structure of the certification body is described in terms of:

- The impartiality it must exercise

- The requirement for it to be an identified legal entity

- Not offering consultancy services

- Structure and competence of the Certification Body's governing board.

Clause 5

This clause places various responsibilities upon the governing board of the certification body.

Clause 6

The organizational structure of the certification body is required to be described and must meet certain minimum criteria.

Clause 7

This clause describes the required competence necessary for certification body personnel:

- There must be documented instructions for personnel describing their duties and responsibilities.

- There must be a selection process to select the audit teams for each assessment by competent personnel.

- There must be certain levels of environmental competence at different levels of the certification body.

- Auditors, as a minimum, should meet the requirements of ISO 14012.

- Auditors should have a minimum specified level of environmental competence, experience and training.

- Composition of audit teams must be appropriate.

- The certification body must have a process to verify that it has the competence to assess an organization within the certification body's scope of accreditation.

- Recruitment and training of personnel are discussed.

Clause 8

Certification bodies must have a system for the control of all documentation relating to the certification system.

Clause 9

Guidance is given for the maintenance of records:

*   How information from the audit itself is relayed back to the certification body, for example

*   Reports by the audit team back to the certification body

*   Nonconformances

*   Time taken over the assessment and the composition of the audit team

Clause 10

Requires that the certification body has facilities and procedures to carry out certification and surveillance of a client's environmental management system.

It also sets out the minimum criteria for implementation of an environmental management system for a certification audit to take place, such as minimum length of time of operation of the system, number of audits performed and a management review performed.

Audit methodology, withdrawal of certificates, the significance of internal audits and regulatory compliance by the implementing organization are laid out.

Guidance on the significance of environmental aspects is also discussed.

Clause 11

Addresses the certification and surveillance facilities required of the certification body – including provision and agreements with subcontracted expertise.

Clause 12

The structure of the quality manual and procedures is set out for the certification body's own management system and how they comply with the criteria of accreditation.

Clause 13

The certification body shall have adequate arrangements to ensure confidentiality of a client's information.

Clause 14

The certification body shall produce and update as necessary a list of certified organizations and make this available to the public.

Clause 15

The certification body will have an appeals process for a ruling on a certification decision.

Clause 16

This is a requirement for the certification body to undertake periodic internal audits.

Clause 17

Guidance for the certification body regarding the use of its registration mark.

Clause 18

This is a directive for all certified organizations to keep records of all complaints relating to the environmental management system.

Clause 19

The certification body shall have documented procedures for the withdrawal and cancellation of its certificates, including dealing with complaints against certified organizations.

Index